Hollywood Outsiders

Commerce and Mass Culture Series
Edited by Justin Wyatt

Hollywood Outsiders: The Adaptation of the Film Industry, 1913–1934
ANNE MOREY

Robert Altman's Subliminal Reality
ROBERT T. SELF

Sex and Money: Feminism and Political Economy in the Media
EILEEN R. MEEHAN AND ELLEN RIORDAN, EDITORS

Directed by Allen Smithee
JEREMY BRADDOCK AND STEPHEN HOCK, EDITORS

Sure Seaters: The Emergence of Art House Cinema
BARBARA WILINSKY

Walter Wanger, Hollywood Independent
MATTHEW BERNSTEIN

Hollywood Goes Shopping
DAVID DESSER AND GARTH S. JOWETT, EDITORS

Screen Style: Fashion and Femininity in 1930s Hollywood
SARAH BERRY

Active Radio: Pacifica's Brash Experiment
JEFF LAND

Hollywood Outsiders

The Adaptation of the
Film Industry, 1913–1934

Anne Morey

Commerce and Mass Culture Series

University of Minnesota Press
Minneapolis · London

Chapter 2 was originally published as "Acting Naturally: Juvenile Series Fiction and the Movies," *Aura Film Studies Journal* 6, no. 2 (2000); reprinted by permission. Portions of chapter 3 originally appeared as "'Have *You* the Power?' The Palmer Photoplay Corporation and the Film Viewer/Author in the 1920s," *Film History* 9, no. 3 (1997): 307–19; copyright 1997 John Libbey and Company; reprinted by permission. Portions of chapter 3 were also published in "The Gendering of Photoplaywrights, 1913–1923," *Tulsa Studies in Women's Literature* 17, no. 1 (spring 1998); copyright 1998 The University of Tulsa; reprinted by permission of the publisher.

Published by the University of Minnesota Press
111 Third Avenue South, Suite 290
Minneapolis, MN 55401-2520
http://www.upress.umn.edu

Library of Congress Cataloging-in-Publication Data

Morey, Anne.
   Hollywood outsiders : the adaptation of the film industry, 1913–1934 / Anne Morey.
      p.   cm. — (Commerce and mass culture series)
   ISBN 0-8166-3732-6 (HC : alk. paper) — ISBN 0-8166-3733-4 (PB : alk. paper)
      1. Motion pictures—United States—History.   I. Title.   II. Series.
   PN1993.5.U6 M6554   2003
   384.8'0973—dc21   2003006494

Printed in the United States of America on acid-free paper

The University of Minnesota is an equal-opportunity educator and employer.

12 11 10 09 08 07 06 05 04 03   10 9 8 7 6 5 4 3 2 1

To Mary Martin McLaughlin
and
to the memory of Pamela Askew
my first and best scholarly models

Contents

Acknowledgments ix

ONE
The Rhetorics of Democracy 1

TWO
Acting Naturally:
Juvenile Series Fiction about Moviemaking 35

THREE
Fashioning the Self to Fashion the Film:
The Case of the Palmer Photoplay Corporation 70

FOUR
"Sermons in Screens":
Denominational Incursions into Hollywood 112

FIVE
Learning to Understand the Foe:
Character Education and Film Appreciation 148

Conclusion 190

Notes 201

Bibliography 213

Index 227

Acknowledgments

This book is based on my doctoral dissertation at the University of Texas at Austin, supported by a Cullen Memorial Trust fellowship. I am grateful to the members of my committee (Phil Barrish, John Downing, Carol MacKay, Tom Schatz, and Janet Staiger), who read this work with great thoughtfulness and insight. I am particularly indebted to my dissertation director, Janet Staiger, who held me to her uncompromisingly high standards. Tom Schatz graciously read chapters more than once, always with a view to reminding me to contemplate the "big picture." I also wish to thank John Belton, Sabrina Barton, Marilynn Olson, and Teya Rosenberg, who read and discussed versions of chapters 2 and 3 with me, and Richard Koszarski, who kindly supplied me with references from his own collection. Librarians who contributed materially to this project include the members of the interlibrary loan department at the Perry-Castañeda Library, Jill Costill at the Indiana State Library (Will Hays Collection), and the staffs of the Billy Rose Theatre Collection and the Rare Books and Manuscripts Division at the New York Public Library. National Board of Review materials are quoted courtesy of National Board of Review of Motion Pictures, Records, Manuscripts, and Archives Division, New York Public Library, Astor, Lenox, and Tilden Foundations. The University of Minnesota Press not only provided a patient and helpful editor, Jennifer Moore, but also sent the manuscript to an anonymous reader whose detailed and formidably knowledgeable comments were invaluable. The Texas A&M University English Department generously supported the preparation of the index. All errors, of course, remain my own unaided work.

Only Claudia Nelson will ever know how much this project owes to her support and encouragement, because my debt to her is greater than can be acknowledged here. She rendered aid both material and spiritual at every crisis, and willingly let herself be harangued on the subject of this book on many occasions. Every author should have such an intelligent and kindly first reader.

CHAPTER ONE

The Rhetorics of Democracy

Art should never try to be popular. The public should try to make itself artistic.
— Oscar Wilde, "The Soul of Man under Socialism"

The 1910s and 1920s witnessed the inception of a particular brand of negotiation between filmdom and its public in the United States. During this period there occurred the creation of a culture of moviegoing that was, at least ostensibly, participatory rather than passive. Hollywood, and its explicators, sought to establish the prospect of a continuum, rather than an unbridgeable separation, between audience and industry—and, perhaps less obviously, between industry and potential institutional critics. Under the right conditions, in other words, Hollywood outsiders could become members of the industry.

Thus, for instance, a screenwriting craze swept America in the 1910s and 1920s, promising riches and fame to any talented or lucky amateur who could produce an acceptable screenplay. Correspondence schools competed to train such writers, while studios and trade periodicals offered prizes for the best amateur offerings. Significantly, this craze (and the schools and prizes) lasted substantially longer than did the period during which studios were genuinely interested in buying products on the freelance market, a period more or less over by 1916. On the one hand, all this attention to a market that no longer really existed was to the advantage of the studios, which benefited from the advertising afforded by the contests, and correspondence schools were happy to have students even if the students never sold their wares. On the other hand, cupidity does not fully account for the persistence of the correspondence schools and allied institutions, because it does not explain how such institutions benefited the consumer. Although the perennial allure of movie glamour was certainly one important factor, what further enticements did involvement in Hollywood offer to the industry outsider?

This question becomes the more compelling when one considers that interest in filmdom extended considerably beyond correspondence schools and their promise to teach a marketable skill. Juvenile series fiction, particularly in the late 1910s and early 1920s, offered narratives about young men and women who entered the film business as actors, writer-directors, camera operators, or exhibitors and consequently repaired their family fortunes; such stories presented the industry as both accessible and beneficent to the struggling outsider. Mainstream religion saw film both as a source of institutional salvation in an era when major denominations were losing membership—Protestant ministers in particular hoped to harness film's power to communicate the Good News and to uplift generally—and as an occasion of sin, since Catholics tended to emphasize the need to make film a "safe" entertainment and to keep members of the Church away from movies that might undermine their faith.

Educators also deemed control of film important to their professional interests. As Lea Jacobs has chronicled in "Reformers and Spectators," by the early 1930s numerous high school courses in film appreciation existed, often teaching students to scorn on aesthetic grounds films that offered "excessive" (and morally dangerous) spectacle and display. Coexisting with film appreciation was "character education," a program designed to use films to make better citizens. Like religion, education took a double-barreled approach to film, defining it as both a danger to be mastered and a vehicle ideal for improving society.

Moreover, both educators and Protestant clergy deemed film's efficiency and allure—attributes that made the medium a hero of the personality age—essential to their own missions. The issue was not the necessity or desirability of using film but rather who could be trusted to shape a film's message, a struggle for control that could affect production directly or operate at the box office. As the Production Code and the Legion of Decency campaign (discussed in chapter 4) demonstrate, those concerned about film content found that they could ultimately shape it most effectively through the establishment of production standards and through box-office leverage; though not all reformers believed in a sovereign public, all had to pay lip service. In other words, even the heated censorship debates of the early 1930s suggested a continuum between audience and industry by making clear to viewers that they had both the power and the obligation to intervene by speaking out to local cinema managers. The latter technique offered the possibility of reformulating the box-office popularity contest, which had always distinguished "good draws" from "flops," into a new expression of the popular will vis-à-vis the film industry: an orchestrated populist revolt against certain aspects of consumer culture.[1] Educational reformers staged a similar revolt in contending that the Production Code itself was too ready to endorse the acquisitive life as the good life.

This book explores the ways film cultures were shaped by forces outside Hollywood, as exemplified by four case studies: juvenile series fiction spanning the period from the mid-1910s through the mid-1930s; the Palmer Photoplay Corporation as a correspondence school of the 1910s and 1920s teaching screenwriting; Catholic and Protestant incursions into Hollywood, ranging from ministerial filmmaking to major interventions such as the Production Code, from the first decade of the twentieth century through the 1930s; and high school film appreciation and character education curricula in the 1930s. As I have already suggested, all four of these case studies employ rhetorics of democracy, an emphasis that raises the following questions: In what ways does the rhetoric of democracy inflect outsiders' interest in the film industry? How do the four sets of writings (in which Hollywood sometimes participated) promote or frustrate various factions' desire to define or control Hollywood as a particular kind of enterprise? How does each type of text define the relationship between the individual and the industry and between the industry and the public? And finally, given an understanding—increasingly prevalent during the period studied—of a constructed or malleable self, what do the texts suggest is the function or mission of the film industry vis-à-vis self-fashioning?

Almost from its beginnings, Hollywood was the subject of a larger cultural discussion that provoked and channeled outsiders' wish to enter and reshape filmdom. While this book makes no claim to be a complete treatment of efforts to reformulate Hollywood to serve particular social or cultural ends, the case studies undertaken afford an opportunity to analyze how four disparate institutions, speaking to constituencies of various ages and degrees of social authority, defined their interest in the film industry. The cases fall naturally into two pairs: not only did juvenile series fiction and the Palmer Photoplay Corporation enjoy their heyday slightly before Christian film reform and high school film appreciation came to the fore, but both also appeared to promise individual social rise through the mastery of some aspect of the film industry; in contrast, denominational and educational incursions into filmmaking styled the appropriate benefits of the industry as more generally social than personal. In other words, juvenile series publishing and the Palmer materials offered the individual filmgoer a method of self-fashioning, while ministers and educators hoped to refashion their institutions via film so as to make the institutional messages more attractive and persuasive. In all four cases, however, close affiliation with film was presumed a mark of popular validation, and in all four one may perceive a belief that film could sell virtually any message, personal or institutional.

Significantly, the texts examined here often addressed audiences who were not merely outsiders to Hollywood, but distanced from the dominant culture as well. Series fiction depicts the triumphs of girls, orphans, the young, and the poor. The Palmer Photoplay Corporation employed women in positions of power and trumpeted the successes of its female students, who purportedly outsold their male rivals, even as it proposed that screenwriting could offer poorly paid men the recognition and wealth they craved. In an era in which anti-Catholic feeling still ran high in many parts of the country, some Catholics used their interaction with film to help establish themselves as community leaders and assimilated "good citizens," and the Protestant clergy who associated themselves with film tended to belong to denominations experiencing a substantial drop in membership. High school film courses molded adolescents, while also promising to give an underpaid group of female and male professionals—teachers—a newly powerful way of influencing society. Moreover, inasmuch as the film industry itself was ostensibly new and flexible, it appeared to offer, for a time, unorthodox routes to success for young people of both sexes. Especially in series fiction and correspondence courses, children and women seemed particularly apt to cross the boundary separating audience from film manufacturer.

But while the consumers of all four sets of writings might have looked to film as a source of power, the writings did not work in identical ways, just as their audiences were not identically disenfranchised. As chapters

2 and 3 will develop, juvenile series fiction and the Palmer Photoplay Corporation emphasized self-commodification and the marketing of affect, promotion of professional skills, and the understanding that individuality would result from internalization of standards derived from contact with the film industry. Through unmediated mutual contact, the individual gained power from Hollywood, which promised to provide opportunities for upward mobility, self-expression, and the development of hidden talents, and conversely the film industry located its own authority in the talented individual.

In contrast, chapters 4 and 5 argue that denominational and educational incursions into Hollywood insisted on controlling contact with the industry, because such mediation was the way to new cultural authority for old institutions. In their desire to reach the mass, reformers were willing to engage in a certain amount of institutional self-commodification, in which clerics and educators experimented with the narrative and sales techniques that were so successful in commercial films. Both clerics and educators also claimed the mass as their ally in their struggle over film content, holding that film ought to be used for public benefit rather than for private gain.

What is ultimately at issue here is that the film industry offered certain individuals the ability to express themselves—in effect, permission to sell themselves directly to the American public—that may have threatened the authority of traditional sources of cultural legitimacy, such as religion, education, and literature, even while indicating newly powerful ways of promoting existing cultural hierarchies. Thus both the lay public and authority figures from other cultural zones attempted to claim the right-of-way in establishing authority over textual production. Over the two decades that this study investigates, the film industry promoted this congestion by basing its own legitimacy alternately on appeals to a public that it claimed to be sovereign at the box office and on affiliation with just those sources of cultural legitimacy that the new medium might be said to threaten.

Each set of writings accordingly deploys a definition of control over "production" that shifts over a relatively small time. Thus juvenile series books, which peaked in the mid- to late 1910s, promise the outsider an active role in all aspects of film production, from acting to shooting to corporate ownership. Correspondence-school manuals, which flourished slightly later, promise access to scriptwriting. The discourse of religion encourages even female censors examining the propriety of certain movie texts in the 1920s to acquire a rudimentary understanding of the industry's economic organization and artistic goals, but not to become studio executives themselves, while it assumes that clergy might enjoy an active role as advisors if not filmmakers. And courses in film appreciation and in character education, the last of this overlapping chronological sequence, narrowed the focus by undertaking only to form an educated film consumer—a consumer comparatively unlikely to make the leap into popular film production, although a Deweyan fixation with "occupation" (discussed in chapter 5) provided some filmmaking experience, typically without vocational intent.

Collectively, the cases describe an arc in which individual, active, profitable participation in the film industry is scaled back to efforts designed to control film interpretation without necessarily gaining or granting direct access to production. Nonetheless, each set of writings proposes that a degree of authorship is vested in the audience through its attendance at films and its exercise of critical judgment. Endorsing one kind of film over another inevitably leads, these discourses claim, to increased production of the preferred type of text, so that audiences are not only "voting" on what they want to see, but even, in a sense, creating it. That this franchise must be presented rhetorically as a fundamentally democratic form of authorship is the leitmotif of each case study.

All four sets of writings examined sought to maintain or promote the notion that the film medium was ideally responsive to public need, that the direction of the industry should reflect public desire, and that the public, through attendance or nonattendance at motion pictures, was conducting a "plebiscite at the box office," as I term it.[2] Chapters 4 and 5 describe reform attempts to redefine this plebiscite into a form of voting behavior that was more republican than democratic, in that direct representation was no longer deemed desirable, but not even the film industry's worst foe wished to attack the fundamental premise that film attendance (or any other way, such as letter-writing, of making one's views as a member of the audience apparent to the film industry) was a form of suffrage.

"Democracy" also had other inflections, such as an understanding that the industrial standards embedded in the film medium made it unusually open to public interest and participation *at an individual level*. For instance, the idea of democracy is latent in the concept of merit rewarded through social rise, an important trope that looms large in the juvenile series fiction and the Palmer Photoplay Corporation materials. In keeping with this notion of individual participation was an understanding that the new medium could make individuals powerful through their control of technology. As Louise Hamburger noted in 1924:

> Spiritual freedom means the birth of power in each individual, the development of his talents, his "differences," the expression of his creative impulses, the subtle stirring of his soul. This can come only through the arts. There can be no real democracy until all can glimpse truth and feel beauty through the medium of some technical expression.
> This is what the moving picture means to the world. (138)

The Problem of the Audience

Great hopes for the film industry aside, however, even critics who were by and large sympathetic to the film industry detected the unfortunate ramifications of the rhetoric of box-office election. Tamar Lane is perhaps the most acute analyst of the motion picture along these lines. In his 1923 treatise *What's Wrong with the Movies?* he argues that

> anyone who has been associated with motion picture production in any capacity knows that every person in the industry is an abject slave to the whims of public taste. Every effort is made at all times to feel the public pulse and prescribe accordingly. In fact, the effort of the industry to give the public what it wants is the main reason for the stereotyped productions which the more intelligent screen followers object to. . . . If the public were to ask for celluloid versions of the history of the recipe of Lydia Pinkham's Vegetable Compound or a film dealing with the life of the man who wrote "Dardanella," the producers would bid one another up into hundreds of thousands to secure the film rights to these masterpieces. (97–98)

In contrast to the optimism of juvenile series books, which suggest that the public en bloc is eager for excellence in motion pictures, Lane implies that audience members divide naturally into two groups, a large group hungry for the trite, and a small group of intelligent viewers whose good taste perennially loses out in the box-office plebiscite. Hollywood, Lane suggests, will always pander to the masses. It was correspondingly the desire of the reformers to enfranchise what they assumed to be the minority, the moral or intellectual elite.

The idea of a public sovereign at the box office becomes especially unsustainable with the acknowledgment of the presumed stupidity of the audience, which did not consort well with a national mood that generally extolled the accomplishments of the expert. The Army intelligence tests administered to recruits and draftees during World War I produced dismay when they revealed that the mental age of the average American soldier was thirteen. Under such circumstances, it became difficult to claim that Hollywood was in the business of making entertainment for grownups if it seemed to be catering to the public. It is possible to claim too much authority for the Army intelligence tests, which proved to have been seriously flawed science; nevertheless, the tests, which purported to collect data on the "innate" intelligence of 1.7 million American men, mark the moment when "scientific" mental measurement became a basic managerial tool. It does not seem excessive to suggest that the tests gave a habitation and a name to certain feelings of pessimism about the future of democracy in the United States, if one believed that the nation would be awash in stupidity either from importing Europe's mental defectives or from breeding its own.[3]

Clearly, as Stephen Jay Gould demonstrates, much of the general rhetoric flowing from the test results stigmatized immigrant and nonwhite populations in the United States (193–233). Despite agitation over immigrant exposure to film earlier in the century, however, the concern over a comparatively low mental age as it circulated in film debates appears to have been considerably more general. It was now necessary, in other words, to worry about the entire audience, which these tests had shown substantially less mature than some observers had hoped. It was even necessary to worry about the censors charged with supervising public morals: in Massachusetts, censors "shall first pass a civil service examination, including the Binet intelligence test for mentality of 18 years" ("Mentality Test," 24). The discovery of the public's potential incapacity made film's acknowledged power more disconcerting, since if the vast majority of the audience was effectively childish, that majority would be disturbingly open to negative influences. Commentators often described these influences as "sex over-stimulation," "sentimentalism," "vulgarity," and the like, making it clear that what was at issue was the purveying of the wrong kind of affect.

Hence commentator after commentator invoked the phenomenon of the low average mental age to explain the shortcomings of an art form that some saw as uniquely responsive to public demand.[4] The unknown author of a *Collier's* article entitled "Why I Am Ashamed of the Movies," who represents himself or herself as a film producer who won't allow his (or her) children to go to most films, sadly notes that "the fourteen-year-old intelligence will be satisfied to pay its way into a show planned by the fifteen-year-olds. The cheap sentimentalities, the vulgarities, the absurdities, will continue to go out to millions a day, making them with each passing week more and more typical movie Americans" (4). The *Annals* of the American Academy of Political Science devoted an entire issue to the status of the motion picture industry in 1926. Arthur Edwin Krows, charged with assessing the relation between "literature and the motion picture," found that the adaptation of literary works with any pretensions to sophistication was made hopeless by the intellectual limitations of some of the audience[5]: "The lowest intellectual level, consequently, is that which governs the character of the appeal to be made. And this is why Adolph Zukor, head of the largest film producing, distributing and exhibiting organization in the world, has publicly found that the average moviegoer intelligence is that of a fourteen-year-old child" (72). Krows suggests that the solution is to develop a distribution system that recognizes the two-tier audience and makes films appropriate to the intellectual abilities of each (72).

Krows's conception of how the film industry should cater to the varying tastes and mental abilities of Americans highlights the rather awkward choices that appeared to face the film industry. It had the option of pursuing a program of uplift that ultimately raised the capacities of the lowest; allowing the quality of production to fall to the lowest common denominator in the interests of box-office responsiveness; creating product appropriate to each intellectual class and station; or relying on standards extrinsic to the audience, imposed by governmental fiat or by self-regulation.

The first choice was more or less continuously popular among certain groups of reformers, who never actually abandoned it (see chapter 5), but it failed to produce results quickly enough to satisfy the film industry's severest critics. The second choice was unacceptable to anyone seriously concerned about the quality of film, because most commentators agreed that what made films injurious was not simply that there were aesthetic lapses, but that the very individuals whose low intelligence made them vulnerable to the worst products of the film industry were precisely those most likely to be already criminal. As psychologist Lewis Terman argued,

> not all criminals are feeble-minded, but all feeble-minded persons are at least potential criminals. That every feeble-minded woman is a potential prostitute would hardly be disputed by anyone. Moral judgment, like business judgment, social judgment, or any other kind of higher thought process, is a function of intelligence. (Quoted in Gould, 181)

The third choice was not successful financially; moreover, the film industry was never genuinely interested in serving a fragmented audience when it could capture the nation by catering to the fourteen-year-old in everyone. The fourth choice is of course the form that regulation of film content ultimately took.

But the rhetoric of the plebiscite at the box office persisted even into the era of the Production Code, because it was so indispensable to the ways film entered American life as an art form accessible to all stations and a business with (initially) comparatively low barriers to entry. These features of Hollywood's reputation, however, were exactly the problem for some commentators made uncomfortable by the nature of the electorate. In 1921–22, when film attendance figures dipped for the first time (a phenomenon that some observers read as a sign of public disaffection rather than as part of a national business downturn), public soul-searching went on in Hollywood. Commentators suggested that if only the "right" people would acknowledge the power and majesty latent in the film industry, they could wrest the product of the industry out of its present low channels and make it a thing worthy of the educated middle class.

"You'll Get What You Ask For," another in the series of Collier's articles by the anonymous producer, pointed out that, as long as the educated middle class failed to go to the movies in numbers, and failed to comment on the aesthetic and moral success or failure of individual productions directly to producers, the average photoplay would wallow in the gutter.

> Theatres, here and there, will find enough good pictures available to enable them to cater, with reasonable profit, to a discriminating clientele, and other theatres, where opportunity offers, will follow suit.

> We shall never get away from a mass of cheap and melodramatic films,
> for there will always be a large class that demands such pictures, but we
> shall have a very different average production, and leadership, in better
> film making, as well. (10)

Significantly, the author of "You'll Get What You Ask For," like Krows, believed that segmentation of the audience (or "classification," as he or she terms it) is where exhibition's future lies. Finally, and crucially, Catholic interest in the Production Code not only endorsed Hollywood's industrial structures but actually built on them to present a single moral standard as not merely suitable but necessary for all audiences. It thus completely sidestepped any notion of audience segmentation based on intellectual or moral capacity.

In practice, some modest market segmentation did take place, but any efficient attempt to serve discrete audiences required two mechanisms not available during the negotiations leading to the establishment of the Production Code, high-quality audience research and exhibitor discretion. George Gallup, as Susan Ohmer has discussed, was the doyen of persuasive audience research (Ohmer, 3), although some statistically based investigation into audience preferences certainly took place prior to Gallup's first investigations of film in 1939 (Maltby, "Sticks," 3).[6] However, as chapter 4 will discuss, the attitudes undergirding audience research were anathema to the Catholic Church, which disdained popularity if it promoted error. The film industry opposed exhibitor discretion because it would have increased production risks—in large part underwritten, under the system of block booking and blind selling, by the independent exhibitor's lack of choice. Richard Maltby reports that his examination of trade papers suggests that *Variety* reviews often contained information addressed to exhibitors suggesting the most suitable audiences for a picture. The most significant gaps in audience construction predictably fell along the urban-rural divide, and typically registered distinctions in "sophistication" among possible audience groupings (Maltby, "Sticks," 6–7). The coming of sound in particular emphasized the gap between audiences at either end of the sophistication scale (Black, *Hollywood*, 56). In the absence of scientific, persuasive audience research and exhibitor choice, Hollywood was obliged to serve the market it wanted—the entire nation.

This unified market was an artifact of the economic structure of the film industry: rising film budgets and production costs were the primary excuses for the industry's need to address the lowest common denominator. The costs attracted the attention not only of religious commentators but also of people outside religion, including one of the most subtle critics of popular culture writing in the early 1920s, Walter Lippmann, who noted that

> If the financial investment in each film and in popular magazines were not so exorbitant as to require instant and widespread popularity, men of spirit and imagination would be able to use the screen and the periodical, as one might dream of their being used, to enlarge and refine, to verify and criticize the repertory of images with which our imaginations work. But, given the present costs, the men who make moving pictures, like the church and the court painters of other ages, must adhere to the stereotypes that they find, or pay the price of frustrating expectation. The stereotypes can be altered, but not in time to guarantee success when the film is released six months from now. (107)

Of course, Lippmann notes here the very economic forces that motivate the textual strategies enshrined in the Production Code: addressing the mass requires reliable, familiar stereotypes with proven box-office pull.

The movie industry, however, is not uniquely guilty of misleading the public; more profoundly, Lippmann finds mass culture as much a symptom as a cause of humanity's detachment from reality. He cites photoplaywright and Columbia screenwriting instructor Frances Taylor Patterson on the necessity of suspense and conflict in the construction of attractive and compelling narratives, a necessity that, he says, applies even to American political discourse (106). Public life, Lippmann argues, is inevitably taking on the coloration of the mass art forms because the latter's practitioners understand the efficient manipulation of the images by which we order our lives. Predictably, Lippmann finds this state of affairs disturbing, and tending toward the undermining of democracy, so that "representative government, either in what is ordinarily called politics, or in industry, cannot be worked successfully, no matter what the basis of election, unless there is an independent, expert organization for making the unseen facts intelligible to those who have to make the decisions" (19). Lippmann does not see the mass media as a thing apart from the rest of society, which can by careful supervision be kept from the moral contagion emanating from film, but rather as a manifestation of all that is wrong with modern society, an outcropping rather than an external threat.

If the mass itself, rather than its entertainment, is the source of corruption, it clearly requires expert guidance, not because it is inherently powerful but because it is insufficiently informed. Lippmann accordingly suggests that public demand for an acceptable technocracy is the solution to humanity's detachment from reality:

> Outside the rather narrow range of our own possible attention, social control depends upon devising standards of living and methods of audit by which the acts of public officials and industrial directors are measured. We cannot ourselves inspire or guide all these acts, as the mystical democrat has always imagined. But we can steadily increase our real control over these acts by insisting that all of them shall be plainly recorded, and their results objectively measured. (197)

Lippmann's emphasis upon a cadre of experts would have resonated (again, in a context that Lippmann himself might not have endorsed) with both attackers and defenders of the status quo in the film industry before the inception of the Production Code. In a criticism that dismisses the concept of an initial outsider becoming a subsequent insider (such as a series-fiction hero or a Palmer graduate), a concept the film industry initially appeared willing to embrace, the anonymous producer of the *Collier's* articles decried the presence in filmmaking of ignorant people who had never demonstrated capacity to excel in any field but motion pictures. In "The Little Men Behind the Big Screen," the problem was posed as the movies so far not having risen above the very low capacities of the pioneering employees, themselves essentially members of the public and thus incapable of providing the necessary uplift:

> The first movie actors were for the most part of the old-time chorus girl and spear-carrier type; the first scenario writers were the shop girls or office boys who were told of the sudden need for stories, with no real training or knowledge of writing—with here and there a newspaper cub or magazine embryo who stumbled into a new gold vein where stories written in an hour could be sold for $15. (12)

Whereas juvenile series fiction and correspondence-school rhetoric validated the talents of just such figures, this purported Hollywood powerhouse dismisses them, implying that the more closed and hierarchical the mature industry can be, and the less accessible, the better for product and audience alike.

But more vital still, in the minds of this critic and others, were the capital behind film production, and the moral values of those putting up that capital. For the anonymous producer, that "the first investors were the clerks or advertising men or born gamblers" is as disturbing as any other aspect of filmmaking (12). Similarly, Lee Hanmer, first chairman of the Motion Picture Producers and Distributors of America (MPPDA) Committee on Public Relations, suggested in 1929 that the happy result of bank and stockholder involvement was that

> financiers realize that it is good business to see that the product of the motion picture companies meets with public favor, and that in the long run the industry will benefit by consistently improving the character of its pictures, even though some of them may not bring as great box-office receipts as the more sensational types. (281)

What's good for the expert—the professional investor, not the marginalized outsider—is good for the country.

For their part, the educational reformers discussed in chapter 5 found within the standardized nature of film a way of establishing an inherently democratic educational structure, as technology would eventually eliminate disparities between one institution and the next through the creation of uniform, high-quality instruction no longer dependent on the vagaries of region or individual teacher preparation. In other words, children all over the nation could be trained to respond to both good and bad movie stimulation as the elite ideally should. The ideal reform approach implied a collective effort between the public and its traditional leaders, in that filmgoers would, by exercising their franchise at the box office, create an alternative form of film authorship.

Authoring, Authority, and Taylorization

The traditional cultural authorities of religion and education thus intervened as much in film as film threatened to intervene in them. Moreover, through its attempts to associate film with other art forms (such as literature or the stage) and to affiliate its mission with that of other institutions (such as education), Hollywood itself established the entrée by which other authorities might intrude into the construction of its product. The public, in other words, was by no means the only authority to which filmdom appealed for legitimacy. Yet series books, Palmer pamphlets, and writings by religious reformers and film educators all acknowledged the moviegoing public as at least potential author as well as consumer of the Hollywood product. Accounts such as Lane's *What's Wrong with the Movies?* may discuss the contributions of directors and writers, but they locate ultimate authority in the audience by ascribing most of a film's character to the manufacturers' responsiveness to box office. While some of this occlusion of authorship arose from the industry's promotion of standardized approaches to scenario writing and of film narratives that appear to "tell themselves," the blurring of authorship within the film industry was a necessary strategy for any group that hoped to dispute the industry's control of its product.

The advantages of constructing the audience as the author of filmic texts were manifold for all parties. Hollywood thereby excused itself from at least some responsibility for what it created; offered itself as helpless to control audience taste (and therefore its own production); and, finally, represented the solution to the problem of morally lax films as one of audience control, rather than external control, of the industry. Reformers likewise found this rhetoric indispensable, because Hollywood's avowed dependence on its box office was also its soft underbelly. Moreover, an extensive discourse invoking the box-office plebiscite was possibly more manipulable than actual referenda on organized censorship: in Massachusetts in 1922, Hollywood won the only referendum over state censorship that was ever put to direct popular vote, implying that direct representation at the polling booth was as unsatisfactory from the reform position as was direct representation at the box office.

The principle of audience as ultimate author also resulted from modes of film manufacture that partook of Taylorization, the use of scientific management strategies, first developed by pioneering industrial-management consultant Frederick Winslow Taylor, that fostered efficiency, often through the separation of the planning and execution stages of work. One of the interesting side effects of labor's burgeoning specialization within American society during the 1910s and 1920s was the occlusion of authorship in a number of endeavors. As Lionel Gossman observes of the new order of industrial capitalism, literature "appear[ed] as essentially different from all other products of labor in the degraded world of industry and the market, but in order to do so, it foregrounded, indeed fetishized, the product, concealing or mystifying the processes of its production" (5–6). Film both glossed over authorship and perversely emphasized what it could not disguise, namely the

technology that lay behind narrative production. As the Palmer materials show, film's basis in manufacturing may have left it the loser in
comparison with more established art forms, but inevitably emphasized
its openness to an interested public. Film narratives, both Palmer writings and juvenile series fiction made clear, required the mastery of technique and the absorption of professional standards. Advancement in
the field reflected merit, not status, and the techniques involved demanded either the expert management of people (in the 1910s a new
professional discipline) or a form of literary engineering that prized an
understanding of how the film medium worked psychologically over
traditional writing talent.

Martha Banta notes that Taylor conceived of his method for stream-
lining production as an ethical as well as a material system, since it had
the potential of maximizing abundance; it laid down the terms upon
which management and labor could converse with each other harmo-
niously (81); and it conflated, in a characteristically American way, efficiency and virtue. These were the very terms in which all four sets of
writings under consideration here explained to the American public the
capacity of film to do good in society. Film, these discourses argued, is
both efficient and compelling, and thus is by far the most powerful tool
for uplift available; however, wide social reach is both a virtue and a
deficiency, depending on the quality of the film.

Film's industrial standardization, in that it had become by the 1920s a
national product designed to reach all markets identically, began to foster a Taylorization of sales.[7] But the industry's opponents used a consciousness of film's sales efficiency as their primary argument for its
control—an attitude fundamental to the Production Code. Indeed, film's
reach highlights one of the most pervasive tensions in reform rhetorics,
the distinction between national and local standards. On the one hand,
the power of a national product to interfere with local mores provoked
a substantial desire to see that product tailored to the concerns of local
populations. If film as a Taylorized industry held out the promise of elevating individuals as far as their talents would take them (Haber, 165)—
a narrative of social rise played out in the juvenile series fiction and the
Palmer materials—consumption made effortless through the Taylorization of distribution and sales offered the contrasting possibility of efficiently debasing a mass. On the other hand, the intervention of reformers
concerned about Hollywood's erosion of local mores promoted na-

tional solutions to local problems and thus tended to move control over film exhibition and even interpretation out of the hands of local groups.

Producing, Controlling, and Selling the Self

This generalization is, however, complicated by reformers' lack of agreement over what to do with the film industry. As I have already suggested and as chapter 4 will detail, Catholics and liberal Protestants differed significantly in attitude and approach. Historians have also noted important divisions of opinion about the choice between self-regulation (preferred by liberals) and censorship (preferred by conservatives) within Protestantism. Generally speaking, Protestants joined Catholics in a desire to mediate the contact between film industry and filmgoing public, and shared, from the early 1920s into the early 1930s, the Catholic anxiety over narratives of sexual self-commodification. As Richard Maltby argues in "'Baby Face' or How Joe Breen Made Barbara Stanwyck Atone for Causing the Wall Street Crash," narratives of women's "self-fetishization," which at times contained an explicitly sexual self-commodification (in which female characters use their sexual allure to achieve mastery), were especially likely to awaken reformers' ire (27–28). But the issue of selling the self extends well beyond questions of sexuality. While it is not always appropriate to link the kinds of self-commodification apparent in the juvenile series fiction and the Palmer discourses to tales about the selling of sex (such as the "fallen woman cycle"), conservative reformers seem to have opposed the outlet for self-expression that the film industry appeared to promise audiences, because they feared that a public sovereign at the box office would encourage the production of increasingly "vulgar" film narratives. Other reformers—those behind the film appreciation movement, for instance—distrusted approaches that smacked of censorship, preferring to educate viewers to reject vulgarity.

Educators, for example, found Hollywood's emphasis on narratives of social rise unrealistic and possibly "undemocratic" in its fixation on "stars" and the labor of the most desirable or visible members of a filmmaking community at the expense of the efforts of the rest. Here educational reformers appeared to be confirming what was, for example, the union criticism of the Taylorist narrative of social advancement, namely that it worked for the rare individual at the expense of the great mass (Banta, 130). This shift in concern, a significant modulation away from the economic individualism that some progressives had praised in

Taylor's approach, resulted in part from the difficulty of promoting such economic individualism during the Depression. It did not reflect an abandonment of interest in liberation through standardization, however, as chapter 5 will detail.

Moreover, while the selling of sex qua sex disturbed religious leaders, the tendency to link desirability and commodification in the 1920s was so pronounced that Protestants found themselves deploying it to retain membership in congregations and rediscover what made religion compelling. Thus Bruce Barton's *The Man Nobody Knows* (1925) argues that Jesus Christ was a virile business leader capable of molding a disparate group of men "from the bottom ranks of business" into history's most effective sales force, "an organization that conquered the world" (Barton, n.p.).[8] To live up to this heritage, the modern religious leader must be able to advertise his or her wares with the most sophisticated and up-to-date devices. Similarly, Protestant reformers were eager to control film not only to ensure that it did not corrupt, but also to harness for uplift what they took to be its essentially compelling nature; this double intent describes not only the earliest Protestant incursions into filmmaking but also later secular refinements such as character education.

Predictably, Protestant, Catholic, and nondenominational reformers insisted that the new film medium must not undermine traditional hierarchies of authority such as those within schools and churches. Instead, they recommended that alert teachers (for example) be quick to capitalize on film's ability to package their messages for wide appeal to the masses. Such mass appeal seemed to some the essence of democracy, while other critics, such as sociologist Donald Young in 1926, worried that this appeal was in fact a fraud on the public. "[S]ocial and economic standards developed by one group," wrote Young,

> may not be adapted to another, may be too expensive for another, and are very likely to be twisted out of shape when another group attempts to adopt them. Yet there are thousands of people in this country, old as well as young, who are consciously and unconsciously trying to live up to standards they have absorbed from other groups as seen in the movies. . . . it is so much easier for the masses to imitate cheaply than to realize the actuality. (147–48)

The commodification of standards, lifestyles, or personality could look like a democratic extension of opportunity to the masses, or like a form of mass-production of the ersatz that would, in Young's formulation, undermine individuality and tradition. Answers to two questions determined which interpretation came uppermost: were the masses ready for the uplift that film promised (materially as well as spiritually), and was the message preached merely one of material self-advancement or did it use film's persuasiveness to argue for morality and good citizenship? In other words, how could film be used to publish the institutional messages emanating from churches and schools while mitigating the influence of narrowly personal stories of getting ahead? Charles Peters's study *Motion Pictures and Standards of Morality* (1933) concluded that film generally both promoted "undesirable" sexual aggressiveness on the part of women *and* emphasized "democratic" values; Peters's findings suggest why reform rhetoric in particular was necessarily so complex.

As far as reformers were concerned, then, the accessibility of the Hollywood *product* constituted the justification for the control of audience desires and personality. The notion of "democracy" manifested here operated in terms more of mass action (reformers uniting to pressure the industry, or designing curricula in order to control interpretation of films) than of individual accomplishment. We may contrast this rhetoric with that of series fiction and correspondence-school publications, which, like other discourses of the 1910s and 1920s analyzed by such scholars as Warren Susman and Stuart Ewen, offered readers the opportunity to stand out by shaping themselves into desirable commodities: actors, successful scenarists, and so on. Series fiction and correspondence schools, in other words, highlighted the accessibility of the industry to *individuals*. Even so, Protestant and nondenominational reformers evidently hoped to achieve a wider audience and a new authority for their messages through an organization-wide version of precisely the process of commodification that the individual undergoes in the earlier, more commercial discourses, which likewise hint that self-commodification (in which the aspirant to a position within the film industry creates a self so fascinating that it can be transmuted into salable fiction) is a means to power over others. All four case studies suggest, however, that the individual or institution seeking to control the film industry had to come to terms with an understanding that internalizing narrative rules and participating in the commodification of the self would have social consequences.

In considering the ways nineteenth- and twentieth-century American cultures diverge, Susman has argued that the earlier period was production oriented but the later period steadily became oriented toward consumption. In Susman's view, further, connected to these two economic modes are correspondingly separate modes of orienting the self, respectively character and personality. While character is inward and stable, and suits the needs of the producer society by permitting or mandating acts of self-sacrifice, personality is a means of self-gratification; personality permits its possessor to stand out from the crowd as he or she seeks to impress others (280). Personality, in other words, is adapted to the consumer society, in part because its traits are presented as those "best developed in leisure time" (280)—an emphasis upon the profitable use of leisure that, as we have seen, educational discourse was developing simultaneously with the orientation toward consumption.

But arguably, the consumption discourse of the 1920s was at its most characteristic when it represented consumption itself as the ideal way to make the self marketable. In the series books, for instance, the protagonists typically attend films as well as making them, commenting all the while on the new appreciation of industry products that they have gained by becoming industry insiders; here, consumption is a necessary form of audience research that will ensure the continued salability of the protagonists' highly personal wares. Similarly, the Palmer materials insinuate that filmgoing is not only a pleasure, but also the means to the improvement of status, the acquisition of skills, or the more mysterious (but desirable) exercise of fascination. To take an example from the correspondence-school literature, we have the case of Frances White Elijah, which illustrates the charm of consumption dressed as the fashioning of the self. She relates in the Palmer Photoplay Corporation organ:

> One afternoon I was comfortably sitting on a country club porch, idly gazing into space and wondering whether it was lack of energy or incentive that was causing me to waste a perfectly good afternoon instead of doing something worth while. As I sat there, an intelligent looking young chap of my acquaintance strolled along. He concluded his greeting with the prosaic question, "What are you doing?" Any one could plainly see that I was waiting for some one to offer me a cup of tea, but I felt the need of giving some kind of startling reply that would make an impression, so I answered: "I am thinking up a new plot for a scenario." It had the desired effect. He dropped into a nearby chair in astonishment. (3)

This vignette confirms Susman's insights while establishing that part of Hollywood's appeal is that it can transmute the lead of ill-spent leisure into the gold of public regard. Here the consumer claims that her conspicuous nonproduction is actually work, a claim that provides her with a moment of mastery over a desirable male. What looks like lolling (significantly, the setting is a country club) is shown to be activity, since the speaker's desire to recoup her lie and become as "astonishing" as she has claimed to be propels her into success as a screenwriter. Series book heroine Ruth Fielding similarly leverages her presence in leisure spaces into a position as the cynosure of all eyes (see chapter 2). Hollywood, then, has the power to make a person interesting by changing years of play into retroactive hard work: gathering material, "thinking up new plots." Indeed, leisure and consumption are mandatory for the would-be photoplaywright, since one of the most constantly reiterated pieces of advice aimed at this group is to see more movies. Even the denominational and educational case studies propose that informed consumption is a social responsibility. Perhaps less obviously, William Lewin, a leader of the film appreciation movement, thought that more efficient classroom instruction was necessary to free children from homework, which might otherwise interfere with family life and other forms of consumption. In presenting parents' views of educational efficiency, Lewin argued that children "must have their afternoons for physical exercise and recreation. They must have their evenings free for music and dancing, for the family round table, for newspapers and magazines, for books other than textbooks" ("Photoplays," 454).

We thus begin to identify the enticements, beyond simple glamour, that involvement in Hollywood offered the industry outsider. The emphasis on self-cultivation ties the discourses in all four case studies to other emanations of the new culture of personality. If advertisements during this period, as Ewen contends, deliberately provoked feelings of unease in the American public in order to sell cures for physical ills, intellectual attributes were by no means exempt from this induced insecurity, an insecurity that certainly extended beyond the rhetorics surrounding film.

Moreover, we may discern in each set of writings a belief that the particular goals of the field in question, whether a matter of getting "personality" across to the multitudes or of publishing an uplifting message, were promoted by film's extraordinary access to the minds of consumers. Understanding how film worked psychologically was the key to success, a fascination that all four cases shared generally with the advertising rhetoric of the 1920s, which manifested a "shift from the 'factory view-point' to concern with 'the mental processes of the consumer'" (Marchand, 11). This concern with the consumer's mentality and probable response to stimuli is exactly what motivated reformers of all stripes to interpose themselves as mediators between the film industry and its audience, even as the same concern was a point of leverage for the individual social rise sought by the protagonists of the series fiction and the students of the Palmer Photoplay Corporation. Contrariwise, when the juvenile series fiction and Palmer materials adopted a language of self-improvement and service to the public, they operated in the zone formerly occupied by ethical and religious systems, and to the extent to which ethical and religious systems preached efficiency and modern psychological methods, they hoped to coopt at least some of the power of the new medium. This mode of framing the discourse implies that it was crucial for reformers to maintain some control over the ways emotional messages were received, since direct control over production was not always possible.

Affect thus becomes critical to controlling consumption: whereas the democratic discourses of the juvenile series fiction and the Palmer Photoplay Corporation indicated that efficient manipulation of affect was bound to be good (since these discourses focused on the deserving manipulator/producer rather than on his or her customers), the more republican discourses of reformism concerned themselves with the naive and persuadable masses. In the view of reformers, those masses, highly vulnerable to outside psychological influences, must either be inoculated against Hollywood's irresponsible use of affect, through proper education, or quarantined with the assistance of censorship.

The Self and the Mass

Clearly the power that the previously unenfranchised attained through participation in commercialized leisure and consumption made some observers anxious. Yet the existence of a consuming mass was by no means seen only as a threat. Contemporaneous theorists of mass society, such as the pioneering public-relations expert Gerald Stanley Lee, saw significant opportunity for the exercise of moral authority in proper leadership of the crowd. In a conception that blends the increasing concern over the maintenance of individuality with the mercantile opportunities represented by the masses, Lee wrote that "a crowd can only be made beautiful by a man who defies it and delights in it at once" (quoted in Bush, 8), and through whose leadership the masses could be led to purchase or vote for the appropriate object or cause. Similarly, Eugene Leach discerns a change in the conception of crowds "from a physically united and active throng" to a "passive body of uprooted individuals," which was all to the good inasmuch as "the mass was initially valued for creating a pliable social solidarity that could be painlessly mastered by progressive elites" (100–1).

The problem with this construction of mass society, however, is that the elites that wish to master it are not, in fact, separate from it. Through laws and institutions designed to pertain to all, elites too could be folded in among the mass, and crowd membership might therefore represent a threat to the self. Thus, in one of the constant ironic tensions of the Progressive Era, systems designed to control crowds and promote social harmony became objects of suspicion. Taylorism, in particular, met with the condemnation of a number of reformers and affiliated sociologists, even though Taylor's goals, such as the elimination of class tension, the achievement of contentment through labor, and the reduction of waste, were all favorite Progressive ideals. The object of reformist concern here seems to be the reliance upon system and the apparent certainty (of Taylor and Henry Ford) that system could to some extent be viewed as coterminous with morality (Banta, 88).

Judith Mayne suggests, however, that group membership may be a source of validation for the individual self, especially selves not otherwise embedded in the traditional public sphere. She sees an analogy between that fragment of the public sphere open to readers in the eighteenth century and the rise of film, given that both are modes of defining community through the consumption of a cultural product. Just as literate women in eighteenth-century England knew themselves part of a network of consumers of such authors as Samuel Richardson, so could twentieth-century women and immigrants perceive themselves as members of an audience for film narratives. In other words, one of the consequences of even such a dispersed public sphere is that its members will see themselves not as isolated individuals but as part of a larger social web—a perception that was one of the functions of the public sphere for more powerful, and politically enfranchised, men. By acceding to the Taylorizing and disciplinary impulses of film as a mass medium that sought to sculpt a particular kind of viewer, filmgoers both gained (or believed that they were gaining) a certain power and were encouraged to take on particular identities. This rhetorical stance informs the Palmer Photoplay Corporation and the religious and educational incursions into the film industry, all of which present as a source of power over the film industry the recognition that their constituents should be considered a "public." Somewhat similarly, Hofstadter suggests that conflict between reformers and the teeming masses they hoped to uplift (in this instance explicitly immigrant) resulted from two irreconcilable notions of public life: the reformer's "idea of political action assumed a popular democracy with widespread participation and eager civic interest.... The immigrant, by contrast... was totally unaccustomed to the active citizen's role. He expected to be acted upon by government, but not to be a political agent himself" (182). While the immigrant may have become a political agent through participation in boss politics, he did so as part of a crowd of similar immigrants who pledged or sold their allegiance, much to the annoyance of reformers, who hoped instead for responsible, individual participation in civic affairs.

These observations reveal the ideology behind the enduring sense of unease with which the politically already enfranchised viewed those newly enfranchised by their participation in the culture of consumption and by later political events such as the Nineteenth Amendment—the comparatively marginalized populations of women, young people, and wage slaves who were particularly the audiences for, and/or the subjects of, the discourses I here examine. As Hofstadter implies, the immigrant appeared to the reformer as part of a crowd or mob, organized perhaps, but not individualized, because of his or her failure to take personal responsibility for political behavior (voting on instructions obviously being anathema to the Yankee ideal). The Catholic intervention in the debates over the Production Code was both a reversal and a confirmation of this view. While Protestant reformers recognized that Catholics were asking for recognition within American public life by assuming a strong stand for morality in film, the approach taken by the Church via the Legion of Decency campaign discussed in chapter 4 may have smacked a little too much of the politics of the ward boss.

To be sure, the prospect of undisciplined crowds participating in cultural events was the object of social concern even before the rise of film. Lawrence Levine reads the progress of the arts during the nineteenth century in America as one in which Shakespeare, symphonic music and opera, and the contents of museums (to take his examples) became successively less "popular" and more "highbrow" at the same time that formerly expressive audiences accustomed to exercising substantial control over performance styles and even content were disciplined into a passive and genteel reception. If, at the beginning of the nineteenth century, Shakespeare's plays were known to most Americans and produced often, even in remote parts of the country, by the end of the century productions had become scarce because the plays had become the preserve of university literature departments, where they were read and not performed.

This transition to a disciplined audience, or even to no audience at all, marked a notable rejection of crowd aesthetics. The cultural phenomenon that Levine describes is analogous to the Yankee ideal in politics; the object, in proper artistic consumption, is not to experience works while part of an undifferentiated mob, but instead to arrive at conditions of reception such that the consumer may experience the work alone, even if in the company of others. Art requires private contemplation, which means withholding applause until the end of the act and not calling for an encore of a catchy aria. While the "disciplined" audience (no longer a throng but a collection of individuals who consume individually) may thus be said to preserve the integrity of the work of art, they are also consuming less actively, at least according to the measure of control *they* exercise over the work (Levine, 194–95). I would contend that film's history as an art form recapitulated the process Levine outlines for a variety of arts in the nineteenth century, only in a much more rapid fashion. In that recapitulation, the distinction between the filmmaker and the filmgoer, at first fluid, flexible, and to some extent easily blurred, became increasingly rigidly enforced—so that by the mid-1930s, outsiders' ability to participate in any meaningful way in Hollywood had become more and more regulated by forces ranging from reform to Hollywood itself.

The Rapt Spectator: Erasing the Boundary between Self and Film

Like many other popular entertainments, film gave critics cause to worry that it was consumed by crowds rather than by discerning individuals. The engrossed spectator, like the politically irresponsible immigrant (and these two figures were occasionally identical in rhetoric), undermined the ideal of public-spirited behavior. Further, filmmakers deliberately

sought to create such an engrossed spectator: the modes of film narration developed by 1916 were designed to ease the connection between
viewer and text, just as the department store had, from the 1880s onward, promoted a similar bond between consumer and goods. Indeed,
1920s advertisers adopted narrative techniques associated with film and
tabloid journalism, precisely to promote such a sense of connection between person and manufactured article (Marchand, 56–63).

These techniques are readily susceptible to analysis by critics. Kristin
Thompson argues that "the classical cinema resulted from a major shift
in assumptions about the relation of spectator to film and the relation
of film's form to its style" (Bordwell et al., 158). She contends that from
the late 1910s onward, Hollywood films minimized the boundaries between a film and its audience, a tack mirrored by noncinematic representations of Hollywood. The classical cinema stressed clarity of narration
to promote maximum absorption. It proffered a notion of the spectator
as rapt, completely enveloped by the performance unfolding before him
or her, and unaware of the physical and psychological gap between performer and audience—exactly the mode of relating to narrative that the
designers of film-appreciation courses in the 1930s sought to discourage
as morally dangerous.

Moreover, if the evolution of textual codes was easing the creation of
an unusually legible mode of communication, the development of the
physical spaces of reception was likewise forming the psychic and cultural structures that would match this narrative development. Miriam
Hansen comments that the spectator is now part of the "film as product, rather than a particular exhibition or show" (23). Hence efforts at
controlling the production of a given film text must ultimately concern
themselves with the project of controlling Hollywood as textual system.

In other words, Hollywood and its critics agree that texts result from
systems, and that it is necessary to define the system to regulate the texts
it produces. Moreover, if the spectator is interpolated into the text, as
Hansen argues, then he or she also results (in part) from that particular
textual system. The juvenile series fiction and the Palmer Photoplay Corporation materials play upon this surmise by offering to reveal the secrets behind the manufacture of film texts, simultaneously presented as
the secrets behind the manufacture of more interesting or more powerful people. Hollywood's textual system, in short, makes the people who
cooperate with it more interesting; it is a textual system designed to

produce "individuals." The religious and educational reformers understand the Hollywood textual system in that they too acknowledge that it
can make its participants powerful, although at times they suggest that
the wrong people are in charge. They redefine this rhetoric into the occasion of intervention by proposing that the Hollywood text of individual
social rise leads to narratives of unseemly self-commodification when
what is really needed are narratives of social benefit, which in turn result from a kind of institutional self-commodification. In short, they view
the Hollywood textual system as a means of producing or reproducing
a society.

Film's role simultaneously as commodity and as hero of the personality age explains why these sets of writings should operate so similarly.
The late nineteenth and early twentieth centuries set up a number of
physical and emotional spaces designed to promote unity between consumer and commodities, as in the case of the department store, whose
mercantile strategy hinted that the distinction between consumer and
product was small and easily overcome. The same lack of distance between consumer and consumed seems part of the textual and exhibition styles of the film industry. Maneuvering the spectator into the text
implies that the normal difficulties in connecting with an artwork can
be overcome by facilitating cooperation with the textual strategies of
that artwork, a process that is a boon or a curse depending on the message. Within this mode of easy connection between consumer and film,
as Young puts it, "We become part of the picture we see, or the picture
becomes a part of us" (147).

Reformism as a Form of Production

But this easy connection between consumer and film is precisely what
reform movements in general wished to frustrate by inserting the reformer into the relay between text and audience. In general, it is possible to say that the ultimately triumphant Catholic approach worked by
insinuating the reform sensibility within the layers of film production
via entities such as the Production Code, in effect writing or rewriting
the text. Educational reformers, on the other hand, attempted to insinuate themselves between text and audience by reconstructing the audience. In attempting to consolidate their control over this authorship,
reformers typically urged structures that interposed some element of
mediation between the public and the film industry, a mediation that

was often to be staffed not merely by the reformers' allies but by the reformers themselves. Thus the Catholic Church found itself supplying some of the Production Code Administration's most important personnel, while Protestant and Catholic reformers sat on film review committees when local censorship was still common. Through film appreciation and character education courses, the educational establishment likewise supplied the manpower necessary to tailor film texts to individual viewers, although it emphasized an understanding of audience preferences more than did the Church.

As I have proposed above, this interest in mediation between film industry and public arose partly from a desire to harness film to messages emanating from educational or religious institutions—but the reformers tended to downplay the mediation, emphasizing the ostensible democracy of their efforts, through appeals to the conscience of the filmgoer, who was encouraged to take responsibility for the consequences of his or her consumption. Indeed, such appeals to the consumer made film attendance a matter of morality and good citizenship. This emphasis upon the "right use of leisure time," to adopt the terminology of the film appreciation movement, marks yet another refinement in the codification, standardization, and supervision of entertainment that occurred in the nineteenth and early twentieth centuries (Butsch, 11–18), and resulted, I would argue, in a more totalizing set of narrative and regulatory structures for the film industry than the film industry alone would have generated.

Given the ease of connection between viewer and film that classical Hollywood cinema offered, some reformers found film a source of concern: a too emulative audience, faced with a too persuasive medium, might be incited to criminality by watching crime pictures, or get a taste for high life from society films. Thus, from the very beginning, censorship or surveillance of working-class amusements was concerned not only with the physical zones of reception or activity (as in the case of the amusement park) but also with the narratives there enacted. Indeed, by approximately 1915 the film industry's desire to upgrade itself to the status of middle-class or family entertainment had largely rendered concern with the physical conditions of reception unnecessary. Nonetheless, a tension between the point of manufacture and the point of sale in the film industry remained, to be exploited by all parties within the shifting alliances of reform, film manufacture, and film exhibition.

The forces of reform were hardly univocal; the initial impetus for improvement in film narratives, for example, was Protestant, but, by the
late 1920s and early 1930s, it was the Catholic Church whose opposition
to the notion of government regulation and whose promotion of self-
censorship on the part of the film industry best matched the industry's
own goals (see, e.g., Couvares, 143). An initial Protestant opposition to
censorship legislation that would have consolidated the artistic "inferi-
ority" of film to newspapers, the theater, and published fiction ultimately revealed rifts in Protestant thought that allowed the apparently
unified Catholic reformers to gain preeminence. In the early years, the
most influential reform group associated with film surveillance was the
National Board of Censorship, which emerged in New York City in 1909
from a prior coalition of the People's Institute and the Women's Municipal League. Members of the Board of Censorship, despite its name,
were actually opposed to formal censorship, and preferred to operate
through a system of endorsement of acceptable pictures rather than
through outright banning of films that failed to live up to their standards (Rosenbloom, "Reform," 53).

The Board of Censorship, later known as the National Board of Review, began a pattern of cooperation, inadvertent or otherwise, between
reform interests and the economic imperatives of the film industry, which
lasted, with shifting personnel, until the formation of the Production
Code Administration (PCA, also known as the Breen Office) and beyond.
The Motion Picture Patents Company (MPPC, an early monopoly of
motion picture producers) and the National Board of Review mutually
reinforced a joint approach to film content until the dissolution of the
MPPC in 1915, a year that also saw the Supreme Court's *Mutual* v. *Ohio*
decision, which stated that, as a commercial enterprise, film had no
First Amendment protection. This decision dismayed not only film
manufacturers but also the Board of Review, whose power faded considerably in the wake of these two events (Rosenbloom, "Reform," 56–59;
Budd, 6).

Maltby and Lea Jacobs have argued that the Production Code itself
operated as a form of generic pressure in shaping film narratives ("Code,"
70; *Wages*, x). The Breen Office succeeded not because it was more draconian than its predecessors but because its employees proved adept at
suggesting the necessary reworkings of narrative and characterization
that would pass the strictures of self-regulation. In almost all instances,

the PCA operated through rewriting rather than through eliminating a production altogether.[9] Thus censorship should be conceptualized as, starting in the late 1920s, exerting some of the force of authorship, from helping to determine which already extant works could be adapted for the screen, to shaping particular treatments of characteristic social problems (such as what Canon Chase delicately referred to as "triangular situations" [quoted in Jowett, *Film*, 174]). Such pressure on studio or author autonomy actually goes back to the days of the National Board of Review, which occasionally proposed the addition of explanatory intertitles to prevent misconstruction by audience members (Rosenbloom, "Reform," 51). That reformers usurped authorial power in this fashion obviously worked to blur the filmmaker/audience divide.

Moreover, some critics hoped to wrest the destiny of motion pictures away from the entertainment world altogether. Chase, for example, sought to redeem film's educational properties, saying:

> The lamentable thing is that the motion picture in the history of the world has not yet been used educationally, scientifically, in our public schools. . . . The remarkable thing is that the motion picture has been seized upon by the amusement agencies of civilization, and the great problem now is how to change that situation. (Quoted in Jowett, *Film*, 170)

Similarly, the Women's Christian Temperance Union actually engaged in filmmaking in the mid-1930s, to produce texts with properly anti-alcohol messages after the repeal of Prohibition (Parker, 75). Thus reformers' desire to manipulate the understandings that films produced existed on a continuum extending from production of films outside the Hollywood system to a set of modifications of Hollywood product invisible to the audience but effectively turning the reformer into a co-author. A desire to make film an educational experience ran a similar gamut, from films produced entirely for this purpose to an attempt to coopt existing films from delight to instruction.

Chase's emphasis on the potential educational properties of film clearly exhibits the "uplift" face of Progressive Era interest, but some reformers (especially those associated with the National Board of Review, such as John Collier) felt that uplift could also be defined as play, and that movies were vital because they allowed for recreation and necessary escape from toil (Rosenbloom, "Defense," 48–51). Although the National Board of Review was interested in educational pictures, it was not at the expense of entertainment (54), which became the attitude enshrined in the Production Code. The dominant mode of censor as author, in other words, established the practice of cooperating with Hollywood over narrative development in exchange for benign neglect of the economic and industrial structures to which the most extreme critics objected. The Production Code accordingly did not satisfy all critics, leading to an attempt (see chapter 5) to establish the viewer, rather than the film text, as the site of a struggle for control. Even here, Hollywood won the battle of the box office as conventionally defined; however, educational reformers hoped to undermine the influence of a widespread national product by the creation of personal standards for the endorsement or rejection of specific film texts. In short, they hoped to exert local control over film attendance on the most microscopic level possible, the individual viewer's mind.

Film and the Public Sphere

Roy Rosenzweig, among others, comments on the apparent identity of interests and ethnic background shared by some nickelodeon owners and their audiences; as the industry developed, however, two strands of changing control emerged. The first is that larger and larger organizations were responsible for controlling theater interests, providing programming, advertising, and so on. This change may have reflected increasing efficiency in production, but it had the effect of moving film programming out of the hands of local audiences (Cohen, 127–29). Second, film narration proper became increasingly the product of national interests. With the inception of the classical Hollywood narrative, circa 1909, film texts grew more and more autonomous and less and less subject to local modification.

This autonomy had numerous ramifications. For one, the local exhibitor had less authority over film content (Musser, *Emergence,* 297). He or she could no longer, for example, determine the order of shots, or supply (once well-developed intertitles arrived) significant parts of the film story line via the spoken word. As the exhibitor was forced out of the role of author, becoming a showperson rather than an editor or narrator, so too was the audience, in that its active participation even within exhibition was increasingly curtailed. These trends reached their endpoint, with nearly complete studio control of textual qualities, with the coming of sound.

Perhaps in compensation, however, film content under the Production Code had to take advantage of subject matter and modes of narration that allowed for multiple interpretations while appearing narratively closed and controlled. Filmmakers had to work out a variety of strategies that divided ways of reading the meanings possible in the narrative: for example, by emphasizing human agency, causing characters to receive appropriate rewards and punishments, mandating that the story perform a prescriptive function, subordinating displays of the body to the story line. At this point, film was assuming a participatory role in society. In other words, film's successful attempt at pleasing the "total picture" school of reformers, those who felt that potentially offensive subject matter could legitimately be presented if properly narrativized, enabled the industry to become a respectable institution of cultural reproduction. Film could, in short, help to shape the self.

Not only individual films, but the institution of film, performed a double function in terms of constituting a site for the creation of a public sphere. Mastery of film texts and intertexts contained the possibility of creating a unified audience through consumption as a universal activity of which the necessity was acknowledged by all. Sumiko Higashi argues that Cecil B. DeMille's fascination with and manipulation of aspects of modern style in his film narratives and mise-en-scène consolidated female audiences in particular, where his films' sexual discourse might otherwise have resulted in a fragmented viewership (160). In effect, the discourse of commodities permitted film narratives access to communities that might have rejected them on moral grounds; the desire for up-to-date information on fashion undermined these objections. Reformers hoping to piggyback on the presumed magnetism of film well understood this phenomenon and thus engaged in institutional self-commodification, hoping, not unreasonably, that their messages would have the same freedom to travel that DeMille's did.

Women were in the vanguard on the new frontier established by consumption, and indeed women appeared to contemporaries simultaneously as chief beneficiaries and most prominent victims of Hollywood. Given the prevailing stereotypes of gender during this period, women's massive presence in the new public sphere may sometimes have made this sphere look like a forum in which participants (conceivably of either gender) were inherently weak-willed and credulous. As Jacobs notes, film appreciation critiqued the feminized and therefore inadequate consumer of popular culture. The feminized consumer arises again in connection with the Palmer Photoplay Corporation, whose materials attempt to downplay film's possible links with more established (and more "masculine") cultural pursuits, even as a corporate spokesperson argued that film should not be seen as the exclusive preserve of female patrons.

Implicit in this understanding of a feminized consumer is a conception of inadequate agency by participants in popular culture. Yet, as some examples above have demonstrated, popular culture did in fact offer new experiences of personal agency, and, what is more, explicitly represented itself as doing so. Thus, for example, Gaylyn Studlar retrieves fan magazines, frequently "dismissed . . . as one more example of women's 'excessive' collusion with the cinematic 'imaginary'" ("Perils," 8), as rather more sophisticated than they are typically perceived to be. She argues that fan magazines played with and anticipated readers' skepticism, acknowledging that audiences would see through obvious strategies of manipulation. The result of these strategies was the placement of the reader "in a discursive relationship of complicity with the stars even while [the fan magazine] simultaneously offered continual demystification of the manufactured nature of the star system" (13). The juvenile series fiction and Palmer Photoplay materials offer similar rewards in slightly different settings.

Studlar's findings suggest that, where Hollywood is concerned, we should constantly look for a tension between rapt absorption in the text and the desire for extratextual knowledge, a tension that arises in different respects in all four of the sets of writings I discuss. All four cases assume that becoming a more educated consumer of film will result in greater self-command and social power. They tell their constituents, in other words, that significant agency is possible, at least rhetorically, within popular culture, although the second two cases shift the field of action away from the personal and toward the social. Each case emphasizes that knowing more about the film industry has both personal and social dimensions, and that the responsible consumer should seek this knowledge.

In short, the film industry both shaped and was shaped by the culture of consumption. It was simultaneously the target of those who hoped to make abundance comfortable and those who rejected certain aspects

of the new culture of consumption. It offered a form of social enfranchisement via access to an alternate public sphere, yet this enfranchisement was never complete or untrammeled, both because film lacked
the First Amendment protections that the traditional political sphere
enjoyed, and because the industry's desire to maximize consumption
mandated cooperation with its critics. Yet industry critics were likewise
unable to have matters all their own way, partly because they too believed
in the power of mass media for personal or institutional self-fashioning
and realized that to have access to these institutions meant cooperation
with them. And finally, the film industry and those who would control
it, like many other businesses during this period, discovered that to have
access to consumers required giving them the illusion of participation,
and possibly something more substantial than illusion.

CHAPTER TWO

Acting Naturally

Juvenile Series Fiction about Moviemaking

> Putting aside men of genius, who may succeed by mere cosmic force,
> your successful man of letters is your skilful tradesman. He thinks
> first and foremost of the markets; when one kind of goods begins to
> go off slackly, he is ready with something new and appetizing.
> —George Gissing, *New Grub Street*

In the 1910s, series fiction for children became a major publishing project for such firms as Grosset and Dunlap, Cupples and Leon, and the World Publishing Company. In particular, Edward Stratemeyer established a long-running, remarkably lucrative publishing enterprise that applied the principles of Taylorism to publishing, as he superintended the mass production of more than one hundred series of books for boys and girls. Stratemeyer certainly did not write every book in every series; rather, he leveraged his understanding of American publishing into a machine for producing fiction in quantity, contracting authors to write according to blueprints. Although this process did not impress critics— Edna Yost, for instance, described it in *Publishers Weekly* in 1933 as "devoid, surely, of either literary sincerity or literary merit" ("Who," 1597)— the books were extremely popular with children. Yost noted that one Stratemeyer series, the Rover Boys, had sold five million copies by Stratemeyer's death in 1930 ("Juveniles," 2405).

Appropriately enough, given his factory-inspired production methods, Stratemeyer was especially interested in technology, an interest reflected in many of his series. Later Stratemeyer protagonists, including the Hardy Boys and Nancy Drew, had no particular technological bent, but earlier series focused on such figures as the Radio Boys, boy inventor Tom Swift, aviators, railroaders, and the various film industry workers on which this chapter will concentrate: the Moving Picture Girls (1914–16, seven volumes), the Moving Picture Boys (1913–22, fifteen volumes), the Motion Picture Chums (1913–16, seven volumes), the Motion Picture Comrades (1917, five volumes), and the Ruth Fielding series (1913–34, thirty volumes), credited respectively to Laura Lee Hope, Victor Appleton (responsible for both the Moving Picture Boys and the Motion Picture Chums), Elmer Tracey Barnes, and Alice B. Emerson.[1] Moreover, in addition to entire series on the new motion picture business, a number of individual books from other, non-Stratemeyer series send their characters temporarily into the world of film, including L. Frank Baum's *Aunt Jane's Nieces Out West,* written under the pseudonym Edith Van Dyne and published by Reilly and Britton in 1914.

This chapter will examine the ways juvenile series fiction characterizes film as a democratic field, and how it shows the hero or heroine becoming materially successful by learning to purvey affect efficiently through a form of self-commodification. This self-commodification not only allows heroes or heroines (particularly the latter) to circulate as branded articles but also requires that they master the standards of the film industry and learn to repress inappropriate feeling, in the interests of creating or controlling feeling in others. Self-commodification has other ramifications, inasmuch as it is a necessary component in the protagonist's relationship with the audience. Here the protagonists are on the receiving end of the buying-as-voting mentality discussed in chapter 1, in which their success as members of the film industry is revealed, maintained, and ratified by the audience response they receive as commodities.

The boy heroes of series fictions operate theaters or become cameramen; the girls act, or, in the case of the omnicompetent Ruth Fielding, act, write, direct, and produce. While all the series offer some information on how films are produced, financed, written, performed, photographed, distributed, or exhibited, the information is typically neither comprehensive nor up to date. Instead, all these series share a crucial set of characteristics, which change but little over time, owing to the inherent conservatism of formula fiction. First, the series show a fascination with the ways film can confound the distinction between truth and fiction, and the details of filmmaking the fictions reveal emphasize film's ability to manipulate conventional understandings of reality by manipulating affect, something they consistently accomplish by substituting the real for the fictional.[2] Second, they stress film's ability to reward plucky and hard-working young outsiders by increasing their social status once they have gained an understanding of the craft. The series protagonist is virtuous but not superhuman, bright and dedicated but no genius, and this very ordinariness encourages reader identification and the understanding that success is potentially available to *any* member of the public who follows a few simple precepts in entering the film industry—an industry essentially democratic in its willingness to open itself to the talented underdog. The underdogs who are the protagonists of these narratives simultaneously undertake to refashion themselves to suit their ultimate role as social and economic success stories. Finally, the stories typically include episodes where the protagonists mingle with the audience to assess the degree to which their work has the necessary appeal. This feature again emphasizes that the protagonists are both our representatives and our servants, consciously and conscientiously seeking to hearken not to their own artistic vision but to ours.

Taken as a group, then, the series books' most substantial function is to explore the ontology of film and filmmaking and to situate it within an industrial context that serves to valorize the new medium by emphasizing its democratic aspects and those of its practitioners. As such, the books have important affinities with the other sets of texts we shall explore, which are likewise fascinated by the commodification of the personal and of affect, the purported "openness" of Hollywood, and the extent to which the film industry is responding to the desires of its constituents.

In addition to the shared preoccupations noted above, the series fictions share a basic plot, whether addressed primarily to boys or to girls, and whether examining the productive or the managerial aspects of filmmaking. The hero or heroine—or, more often, a group of heroes or heroines—is an emancipated adolescent or young adult, frequently orphaned (Ruth Fielding) or emerging from straitened financial circumstances (the De Vere girls in the Moving Picture Girls series, Frank Durham of the Motion Picture Chums); thus the protagonist has more authority over his or her development than has the average teenager.[3] Moviemaking represents a fresh field for the correction of the family fortunes, and one that seems especially welcoming to youth and even femininity. While the age of the typical Palmer Photoplay Corporation student is greater, Palmer's vision of the talented but downtrodden outsider whose dreary life may be transformed by entry into the film industry is strikingly similar, as chapter 3 of this book will detail.

The young heroes frequently are befriended by one or two adults sometimes less shrewd or competent than themselves. The protagonists thus have to ascend professionally and socially alone or with the companionship (and sometimes the material aid) of their young friends; adults are convenient but not essential. The dramatic tension in these works always emerges from the machinations of the protagonists' business rivals, who will stop at nothing to deprive them of their ideas and their success in the marketplace. Typically, plots revolve around theft of money, personnel, locations, ideas, patent rights, props, scripts, equipment, and even reputation. Even those books that, like the Motion Picture Chums series, focus on the managerial side of film exhibition suggest that the heroes are subject to thefts of *intellectual* property—namely, their good exhibition ideas, their understanding of marketplace dynamics, and their trained assistants (who are vulnerable to abduction). The antagonists are also willing to stoop to sabotage, some of it potentially lethal. Needless to say, the protagonists triumph owing to their superior understanding of the marketplace, their grit, and their dedication, as well as their detective ability. A common secondary plot strand is the protagonists' battle with the forces of nature, which at times work in concert with, and at other times work against, the antagonists. The protagonists almost always exhibit their generosity by rescuing even their enemies from disaster.

Through the protagonists' success, series fiction demonstrates that even children can wield authority within the new medium. Indeed, the novels suggest that these malleable, hard-working, and upright children are arguably the best authorities in this brave new world because they are willing to refashion themselves to fit its demands, and because their very presence in the industry tends to assure an industry as pure as the heroes themselves. The film industry's essential democracy emerges through the construction of working-class boy or penniless orphan rising to comfort or even prominence through participation in it. Specifically, protagonists achieve this prominence through the development and sale of affect, coupled with their ability to repress their own feelings to play upon the feelings of others or to perform filmmaking labor. Thus, series books represent one way the path to success in the film industry is mapped for the outsider: one wins out by controlling one's own anxieties or desires (for leisure, for instance) and by satisfying the public's need for ever-fresh spectacle, an enterprise that often entails self-commodification as protagonists serve up their own traumatic experiences for public consumption. A system of democratic exchange thus operates, in which individuals offer themselves to the larger audience and in return gain power and status; since, however, the public is what ultimately confers these prizes, the public may ultimately reign supreme through the election and anointing of its representative.

The Algeresque Matrix

Series fiction manifests a certain ambivalence toward both technology and capitalism. Film is at once the field for new struggles over technology and money and the remedy for old contests. John Parris Springer rightly observes the similarities between Horatio Alger's accounts of upward mobility and the representations of poor children making good in these series fictions; the connection is only too predictable when one considers that Stratemeyer began his career at Street and Smith by finishing incomplete stories by Alger and William T. Adams (creator of the popular nineteenth-century "Oliver Optic" tales) for publication (Billman, 19). But just as Michael Moon has suggested that Alger's apparently straightforward narratives mask complexities of gender and sexuality (89–90), the series books' ideological positions, and the convergences and divergences between boys' series and girls' series, are rather more complex than at first may appear.

It is worth pausing to examine some aspects of the Alger paradigm, because its conventions shed light on the construction and function of the Stratemeyer offshoots. While one typically remembers Alger's heroes rising through society on the basis of carefully cultivated virtues, Marcus Klein notes that "the practice of the virtues was never the proximate cause of the rising. In Alger's arithmetic of luck and pluck, the former always greatly outweighed the latter" (23). This emphasis on luck—theoretically available to all, of course—is something that the Stratemeyer books preserve, inasmuch as the heroes and heroines often get their start through the performance of a rescue or favor that has nothing to do with the profession to be later pursued. Ruth Fielding gets her first small capital not from screenwriting but by receiving a reward for finding lost property (*Ruth Fielding and the Gypsies* [1915]),

and the Motion Picture Chums get their start in exhibition by prevent-
ing a runaway carriage from causing serious damage (*The Motion Picture Chums' First Venture* [1913]).

The prominence of chance correspondingly diminishes the power of
the individual agent, an aspect of the series books important for its relationship to the rhetoric of the perfect accessibility of the film medium
to all comers. Indeed, the lottery aspect of success is something that
Theodor Adorno and Max Horkheimer comment on with reference to
radio and film, arguing that it tends to obscure the unequal nature of
access to success within media cultures:

> The starlet is meant to symbolize the typist in such a way that the
> splendid evening dress seems meant for the actress as distinct from the
> real girl. The girls in the audience not only feel that they could be on the
> screen, but realize the great gulf separating them from it. Only one girl
> can draw the lucky ticket, only one man can win the prize, and if,
> mathematically, all have the same chance, yet this is so infinitesimal
> for each one that he or she will do best to write it off and rejoice in the
> other's success, which might just as well have been his or hers, and
> somehow never is. (145)

And to be sure, if success in a new field is open only through the con-
tacts gained from unplanned encounters, then it is not possible to rise
through a frontal assault on the bastions of capitalism. Yet success by
lottery suggests at least that no one has imposed systematic structural
barriers against social rise by individuals of the wrong age, class, or gender. Virtue and good fortune are both necessary, although Adorno and
Horkheimer might argue that the media are using luck to suggest that
virtue or skill is rewarded more systematically than perhaps it is; even
so, the lottery aspect that these critics so decry appears to owe as much
to the form of the series narrative as it does to outside rhetoric about
the film industry.

Where the Alger tale seems more modest than the Stratemeyer series
book is in suggesting that the rewards that follow from the joint operation of pluck and luck have limits. As Moon puts it, "'Rising' for Alger's
heroes always remains a waiting game; within this pervasive passivity,
there is an active and a passive position, but there is no way for a boy to
take a more direct approach to the world of work and achievement in
Alger's books" (89). Moon's observation is basic to his argument that
Alger's works do not describe a world in which a nobody rises from the

gutter to the boardroom, but rather an environment in which an already promising lad of character and intelligence rises to a comparatively modest position in the capitalist establishment (89). Significantly, the earlier Stratemeyer series concerned with technology or acting tend to conform to this paradigm, in which the successful young people occupy positions valued but subordinate to those occupied by more senior managers in their film production companies, with subsequent series offering protagonists greater power. Even within the Stratemeyer series, in other words, we have a split between the earlier, more modest Algeresque narratives of rise to a safe if subordinate position and the later narratives of rise to greater prominence.

The exceptions to the Stratemeyer norm of modest success are the Motion Picture Chums and Ruth Fielding. The Chums' success is comparatively plausible, but in the concluding volumes of the latter series, Ruth has reached the top rungs of the film industry in most of its significant branches; she is simultaneously actress, scenarist, producer, director, and studio owner. Further, the jealousy her success has always inspired in the less deserving has reached such a pitch that her physical safety is frequently at risk. Notably, the Ruth Fielding series was the only one to continue into the 1930s (its closest rival having ended by 1922). Ruth's greater longevity may explain why this series appears to participate more completely in the twentieth-century culture of celebrity by emphasizing the greater rewards accruing to a position within the film industry: the potential profitability of work within Hollywood was much more obvious in the 1920s than earlier. We may also note that the promise of spectacular success lifting one far above the herd was typical not only of the Ruth Fielding series but also of the discourses surrounding the Palmer Photoplay Corporation.

One function of both Alger and Stratemeyer narratives, as critics have noted of the Alger story, was to provide a kind of map or encyclopedia of the range of new urban experiences available to readers, at least vicariously. Not only does the typical Alger book describe the many kinds of confidence tricksters dotting the nineteenth-century urban landscape, it also offers a significant collection of information on more mundane topics, such as the topography of New York City (Nackenoff, 42–44, 49; Klein, ch. 2). That the Stratemeyer series books (like non-Stratemeyer texts such as *Aunt Jane's Nieces Out West*) painlessly purvey industrial facts enables them to operate in a way analogous to fan magazines and other sources of information about the film industry, as both Kathryn Fuller, who wishes to link the series to the hobbyist discourse that shaped some writing about film during its first fifteen years (125), and Sherrie Inness, who feels that series fiction made the new film technology more accessible to girls in particular (173), have argued.

Affect and Social Rise

Moon suggests that the Alger series is potentially most informative when it explores basic labor/capital relations (89), a statement also true of the series fiction we are considering. The provision of up-to-date, detailed technical information is not the real strength of the Stratemeyer series. Rather, they excel at providing a particular context for film labor by presenting the ways it is open to the potentially disenfranchised (girls, teenagers, ailing actors, orphans, and the like). Filmmaking, these books say, has properties that make it superior to other occupations, such as working in factories.[4] Not only can one rise further in the film world than through assembly-line work, one will also have more mobility, greater scope for creativity, and more control over what one serves up to the public. But the series books' fascination with the relations between film and truth suggests that insofar as they are concerned with explicating the mechanisms of social rise through involvement in the film industry, they are really demonstrating how to turn feeling into power, a transformation at which filmmaking proves particularly adept, and a rhetoric shared with entities such as the Palmer Photoplay Corporation. Entering the film business allows one to package and sell others' feelings (or one's own); that fiction keeps turning into fact (as when, e.g., a simulated prairie fire becomes a real threat to the protagonists) assures the reader that a fictional film similarly depends at bottom on real emotion. Since feeling is primarily what films have to offer, both customer satisfaction and retailer success require that the goods be genuine, a trade secret that these books deem considerably more important than the mere recounting of facts about filmmaking or the film industry.

In these texts, filmmaking, not yet seen as a glamour industry capable of propelling its exegetes to fabulous wealth, is shown to foster middle-class success. Filmmaking's very accessibility to children means that the formerly less productive members of the family can now be wage earners. This association with domesticity renders filmmaking "safe." Alger heroes such as Ragged Dick are made appealing in part by their eagerness to construct substitute families for themselves; similarly, there is more appeal than threat to us in the autonomy of the Moving Picture Boys, say, as unmarried and originally parentless boys who can and do go anywhere in the world with no adult supervision, but use their freedom to find Joe's long-lost father and sister and settle down as a happy ménage à quatre "in one of the New Jersey suburbs" (Appleton, *Earthquake*, 18).[5] As the servant of the public, the filmmakers must participate in public mores. Yet, like any other industrial activity, filmmaking has its share of danger, which is another manifestation of its accessibility: if children can enter it, so can criminals. This danger, of course, is necessary for the books' own production of affect, in that it furnishes excitement for the plots. But the danger contained in the books' adventures also tends to ratify the affect produced by the films within the books. Both sets of narratives tend to be the same kinds of stories, which produces a corollary to the observation about film's accessibility: if children can have dangerous adventures in any context, they can enter the film industry.

The adventures that Stratemeyer's children undergo are designed to produce and exploit affect in both the reader and the child hero. Moreover, the adventure plots tend to depend upon film's ability to transcribe reality. On the one hand, as a documentary medium of new persuasiveness and immediacy, film can be a tool for righting social wrongs; many books contain scenes in which the heroes catch thieves and cheats in the act by exposing their hypocrisy on film. On the other hand, film's connection to reality is potentially frightening because it allows the capturing of the sufferings of actors (and others) as part of either a fiction or a nonfiction narrative, and the two genres tend to coalesce. Thus many of the books rely upon events initially staged to create fiction films, such as fake floods, fires, cave-ins, and runaway animals. With astonishing frequency, and often through the mischief of professional rivals, these staged events escape the control of the film crews; consequently, the camera operators and the actors must confront the dilemma of whether or not to record actual misfortune. Does it matter whether the misfortune falls only to the lot of the professional paid to undertake it? Is it acceptable to photograph death and pain? While the books do not offer particularly sophisticated answers to these questions, they raise them in a way that suggests that technology is more problematic than not, and problematic precisely because of its status as the recorder, purveyor, and potential creator of emotion.

Moreover, the conflation of fact and fiction, actual and staged, ultimately leads to difficulties in defining film work as labor. Earlier narratives in particular praise the parsimony of film companies that turn genuine disaster (a fire, a storm, an earthquake) into effective fiction. The ability to transmute misfortune into displays of authentic emotion appears the sine qua non of filmmaking even as late as *Ruth Fielding in Talking Pictures* (1930), when Ruth directs by telling an untried and apparently mediocre actress, "Now, Miss Calvin . . . forget about yourself and think of the story. Your friend has just brought you bad news. Register surprise and pain." When Edith twice fails at this task, Ruth instructs her cameraman to continue filming after Edith is allowed to think that she has been dismissed, thereby producing the best possible facial expression for this scene: "For the first time Edith looked into the camera and gave a self-conscious start. For a moment she did not understand, then it slowly dawned upon her that Ruth had ingeniously forced her to act naturally" (79–80). Ruth's technique consists of substituting the genuine for the inadequately fabricated emotion. This slippage between real emotion and an acceptable performance constantly troubles both the film plots and the narratives of the novels in which they are embedded.

Capitalism and the Family

Another theme that runs strongly through every book in every series (and Baum's *Aunt Jane's Nieces Out West*) is that of the business rival who will stop at nothing to secure the ruin of the protagonist(s). The ongoing emphasis upon both the film medium's and the film industry's ability to restore family fortunes or to right wrongs clearly stems from the conventions of melodrama, a mode that pervades every aspect of most books within these series. But the business-rival motif not only functions to keep suspense alive; it also works to distinguish good business practices from bad. Moreover, the emphasis on commercial matters is, in another reference to nineteenth-century melodramatic tropes, always bound up with the potential construction or destruction of family. Thus if film technology represents a solution to economic and social problems, the cutthroat world into which it invites the young heroes and their families is one in which only the most extreme self-discipline and fortitude can hope to prevail. The world of film entrepreneurship, these books suggest, is so exciting and lucrative that it positively invites mayhem. Two factors are at work in giving the film industry this character: its ability, through the kinds of self-cultivation it promotes, to create the fascinating personality or lifestyle that is the envy of all, and the purported ease of obtaining an entry-level job in this business.

Capitalism, in other words, has two faces: a kind one, represented by the family to be made with like-minded friends (as in the all-male "family" of the Motion Picture Chums) and thoughtful employers (like Mr. Pertell of the Moving Picture Girls series), and a brutal one, embodied by the antagonists whose success mandates unfair attacks on hardworking young people. Frequently, the films that the protagonists make are themselves narratives about the good and bad faces of capitalism, particularly in the earlier books. For example, in *The Moving Picture Girls at Rocky Ranch* (1914), Alice and Ruth DeVere appear in a film about a shirtwaist factory, shot on location. When a real fire breaks out at the factory, the "good" conditions of capitalism are apparent in the deportment of the female manager, who sees that all operatives get out safely, including the actresses. The manager's firmness and heroism give the lie to the comment made by one of the less sympathetic members of the acting troupe, who remarks, "This is a regular fire trap! All shirt waist factories are" (42). This narrative of manufacturing conditions for young women stands in pointed contrast to another story fresh in the minds of readers, namely the infamous Triangle Shirtwaist Company fire of 1911, which killed 165 sweatshop employees and evidently inspired at least one real-life film during this period (Ross, 34–35). Similarly, in *Aunt Jane's Nieces Out West,* Beth and Patsy's interest in motion pictures is kindled through the revelation that one can use film to expose the abuses of industrialists, who erect shoddy factory buildings that might collapse on their workers; to further the girls' new connection with Hollywood, they meet the philanthropic capitalist A. Jones, a large investor in the film business who offers to spend an additional 1.4 million dollars to build a chain of movie theaters to promulgate healthy entertainment for the young.

In the case of the interpolated narrative in Baum's story, the selfish industrialist suffers because the collapsing wall crushes his own child, a plot twist suggesting that what capitalism primarily attacks is the family. Conversely, association with the film business in series fiction represents an important antidote to the rigors of industrial behavior because it allows both the development of acceptable surrogate families and the maintenance or reconstitution of biological families. To state the case another way, the series narratives propose a blending of work and domesticity that permits a seamless fit between these two spheres. The corporate self imaged by such fictions, in which heroes travel in packs rather than brooding Byronically apart, permits wage-earning and family togetherness to be synonymous, just as Alger's Ragged Dick offers a home to a younger boy who can tutor him in refined ways and turn their shared room into a source of upward mobility. The neo-Algerian world of film described by the series narrative figures as desirable because it simultaneously makes family life exciting and profitable and work warmly domestic.

The Motion Picture Chums may be the most extensive example of a surrogate family arising from the association of individual members with a facet of the film industry. While some of the Chums come from middle-class homes, others, such as Dave Sawyer, whose father is an unreliable drunk, are escaping victimization at the hands of their former employers, themselves affiliated with the motion picture exhibition business (*The Motion Picture Chums' Outdoor Exhibition* [1914]). The Chums form the ideal family, and, in these most Algeresque of all the series books, the heroes' success is largely attributable to their superior understanding of exhibition's essentially domestic nature. As the narrator observes of their flagship theater, the Empire, "It was like getting back home to once more enjoy its coziness" (Appleton, *Idea,* 14).

Thus the Chums succeed in their first venture because they are neat, clean, and aware of what attracts patrons (Appleton, *Venture,* 104), while their rivals hope to spoil their success by destroying the quiet and order necessary for successful exhibition. They also succeed because they understand the importance of home to their audiences; in the same novel, they help a farmer by flashing a slide during the film urging him to go home to put out a fire. Finally, they understand the Progressive zeal behind the regulations governing exhibition as a kind of internalized set of housekeeping rules, and thus their instincts are better than those of their adult competitors: "The [rival] place looked clean, but was poorly ventilated. There had not been much attempt made at ornamentation. The auditorium looked barn-like on account of its great width" (*Idea,* 149). As if this were not enough, the boys notice that the theater lacks adequate fire exits, suggesting that they, and not their adult rivals, have mastered the elements of Progressive Era reform of public entertainment venues. The presence of children in managerial roles firmly indicates that child protection need not be an issue when children are themselves in charge. In short, the Motion Picture Chums fashion themselves into an ideal family by uniting the manly qualities and business instincts of the Alger hero with the domestic virtues required of a good housekeeper.[6]

On the other hand, the involvement of the Moving Picture Girls and Ruth Fielding in the film industry tends, if anything, to vitiate certain conventional aspects of domesticity for these heroines, who like the boys must always be ready for adventure and take their knocks in the name of art or truth. For example, the Moving Picture Girls are rarely at home, because they are often on location. Even when the girls are shooting in New York, where they live, the texts tend to associate the markers of domesticity with the new virtues of consumption, such as buying new clothes for shoots and eating in restaurants, rather than with the old wifely virtues of home manufacturing. Indeed, the girls must consume clothes, trips, meals, and the like to be productive actresses, since exciting narratives require that readers see them enjoying or experiencing these things; the melodrama of the films they make obliges the consumption and destruction of goods, which are, in effect, rendered into salable experience.

While the girls are willing to simulate more conventional feminine employment, the novels (unlike the films in which the young actresses star) make clear that this behavior is simulation rather than actuality. Take the occasion upon which Alice and Ruth DeVere play sweated laborers in the shirtwaist factory when all they know of shirtwaists is how to choose them: "'You don't really have to sew,' explained Mr. Pertell. 'It is all machine work, anyhow,'" a comment that critics might levy against the film medium itself (Hope, *Ranch*, 38). Similarly, the girls are content to pursue family life solely within the confines of their work family, which consists of themselves; their father; Mr. Pertell, who is a more powerful father figure, as their director and producer, than their own father is; and all the other employees of the Comet Film Company, including Ruth's cameraman beau, Russ. Indeed, the acting company, says Mr. Pertell, is "as nearly like a happy family as I can [make it]" (Hope, *War Plays*, 51). Like the shirtwaist script, this surrogate family often obliges the protagonists to simulate a more conventional domesticity than they experience offscreen, while at the same time it enables them to keep their biological family together even though their father, an actor who has lost his voice, can only find employment in the world of silent film.

In short, the actresses engage in derring-do in manufacturing films, but often in aid of portrayals of nineteenth-century and rural life. They know themselves to be modern women, performing in a modern idiom, and believe that "Girls ought to be brought up able to do something" (Hope, *Girls*, 54), but the narratives they appear in tend, on the whole, to be *retardetaire*. Such devices inform the audience that film work can free one from the demands of convention, but only because it involves the simulation of obedience to traditional social mores. Anxious readers receive reassurance that, although film looks bohemian, it does not seek to undermine tradition; less conservative readers may see that by substituting for real life, film can liberate its workers to develop their personalities in new and potentially more desirable directions.

Or take Ruth Fielding, whose social rise is actually a return "home" to the middle class, from which the death of her parents temporarily banishes her. By performing services for ostensibly more powerful figures, she earns enough to go to boarding school and then to college, so that she can continue to acquire the attainments of a privileged young woman and to associate with her friends Helen and Tom Cameron, who travel in more exalted circles than would Ruth without this small capital. Through her work, Ruth earns her way to the kind of family she wants and deserves. Getting ahead, however, means never enjoying leisure— or constructing a domestic nest—in a conventional fashion; Ruth works ceaselessly at her profession while Helen and Tom, her future husband, urge her to take life more slowly. Despite their comments, one way we know that Ruth has risen is in her inability to stop working. She is a manifestation of what the sociologists Robert and Helen Lynd notice about the business class in *Middletown in Transition* (1937): using leisure to get ahead results in never having leisure at all (244).

Screenwriting is thus the ideal vehicle for Ruth's social rise (and for that of the clients of the Palmer Photoplay Corporation, who are described as thinking out stories at home or on the veranda of the local country club), because one can perform it at any time and in venues not normally associated with work. More specifically, it is a form of labor that will never declass or desex her, because it does not appear to behave like conventional labor for sale, and it does not separate her from her more privileged companions, because she can undertake it in settings, such as lakeside resorts, boarding schools, colleges, and exotic countries, that are the haunts more normally of the wealthy than of friendless orphans. Ruth can thus use her work to establish a new family, especially since, after she becomes affiliated with a production company and has some control over casting, she is able to lift friends from obscurity or inferior labor into higher status.

On the other hand, boys' work in these series tends to associate the protagonists with the most tangible aspects of film production, such as the conversion of buildings into theaters or the shooting of film. Physical artifacts, in other words, result from their labors, which may reflect the presumed greater male affinity with manufacturing. Ruth's fiancé, Tom, manages the finances of the studio rather than serving as another star director or performer. Boys' labor also involves tending machinery (the cameras), a job that, because it results in the production of things (the finished films) is more readily definable as work and appears less dependent for its value upon subtle changes within the self. Writing, acting, directing, and producing, which may be harder to define as labor because they are new fields in the motion picture context, fall within the feminine sphere.

Girls' acting and writing in particular is a form of self-cultivation; the ultimate value of their work lies in the manufacture of an inherently more desirable persona, both through the development of skill and through becoming better known, which efforts tend toward an increase in "fascination." The books insist that girls perform work, but tend to elide the actual technique that might define acting and writing, in particular, as labor. While both boys and girls use film to fashion themselves—to leave their originally more limited walks of life—the girls are using their work to manifest or shape a new self, which they may then market for the entertainment of the public and for their own social rise. Typically, boys such as Russ (the Moving Picture Girls' cameraman) and Tom labor to free up their womenfolk for this commodified self-cultivation.

Leo Lowenthal has noted, in the case of biographical profiles published in American magazines between 1901 and 1941, a shift from subjects who function as "idols of production" to those who function as "idols of consumption"; the former live out Algeresque dreams of working a way upward, while the latter model ways of using leisure time (quoted in Dyer, 45). As Richard Dyer observes, Hollywood fan magazines began in the 1920s to suppress the notion "that making films is work, that films are produced" (44). The series books, most of which antedate this particular glorification of leisure, suggest that the latter may be rooted in gender-role distinctions that become less obvious in the post–World War I era. But while later fan discourse would be constructed to imply that work in Hollywood is really leisure, series fiction seems anxious to make women's employment in film more worthy of respect by pointing out that although dressing up, emoting, and improving the self may look like leisure activities, and are all the more appropriate for women on this account, they are really work.

Self-Cultivation, Self-Revelation, and Audience Validation

Precisely because Ruth's screenwriting seems so powerfully directed toward creating the persona and family that she wants to inhabit, it becomes all the more necessary to emphasize the effort inherent in that screenwriting. Ruth must repeatedly try to convince her great-uncle, a miserly New England miller, that her labor will actually result in material success. If one measures success in strictly financial terms, the books make clear that Ruth works hard and earns plenty of money. But in other respects, the novels as a group have some difficulty in getting across what sort of profession screenwriting really is or how it functions as labor. Indeed, most juvenile series fiction about the film industry inadvertently makes difficult the depiction of filmmaking as labor for two reasons: filmmakers, whether Ruth Fielding or those indefatigable documentarians the Moving Picture Boys, succeed best when their work is the truest depiction of a reality unfolding before the camera, and success in the marketplace goes to those who are there first with a novelty. This approach suggests, on the one hand, that there is no such thing as a completely original fiction (because many film narratives are traceable to some opportune occurrence that the filmmaker catches on the fly) and, on the other, that the definition of originality in the marketplace is so stringent that there can be only one member of any category before the public will tire of it.

To the extent that a film must mirror presumed reality, the filmmaker's contribution is largely technological or managerial rather than creative. Even though Ruth is ostensibly writing fictional scenarios, the mystery or adventure device at the heart of each novel often undercuts the presentation of her work as wholly original. First, the mystery functions to elide the routine of filmmaking labor; far from being a daily grind, work becomes a series of life- or reputation-threatening problems. By the standards of men such as Ruth's great-uncle, excitement and "work" are antithetical; if Ruth's job entails both the construction and the experiencing of melodrama, can one really term it a "job"? Moreover, since the mystery frequently results in Ruth and her assistants actually experiencing the events she has planned as fiction (just as the Moving Picture Boys must experience the nonfiction events they chronicle), her films are not so much performed as realized. Technique notwithstanding, it would appear that filmmakers have difficulty in fabricating experience; they must always re-create it instead.

This experience may involve either others or the self. While the boys tend to employ film as an external truth-telling mechanism, ideal for catching malefactors, the girls find that film can also offer a forum for self-revelation. For example, Blake and Joe use their prowess as cameramen to reveal the guilt of a hit-and-run driver, a saboteur, and a German spy (*The Moving Picture Boys* [1913], *The Moving Picture Boys at Panama* [1915], *The Moving Picture Boys on the War Front* [1918]); similarly, in *The Moving Picture Girls Snowbound* (1914), the girls and their cameraman friend Russ are able to prove insurance fraud when they film a supposedly injured man performing heavy labor (205). Significantly, however, the role of revealer of external truths typically belongs to the young man who operates the camera; a more personal revelation appears to be the province of the female actor. Moreover, the females' exposure to film apparently holds out the promise of greater self-expression, even in areas outside the realm of acting. For example, in *The Moving Picture Girls at Rocky Ranch*, Alice is notably fond of slang, for which her more staid older sister, Ruth, reproves her. Alice shifts the blame for her choice of words to their new profession, saying, "It's this moving picture business. It just makes you use words that *mean* something, and not those that are merely sign-posts" (1–2). Slang, like the girls' frequent performances of genuine fear and distress, suggests the immediate expression of sensibility, and, indeed, the economic and social status that the girls derive from their profession. (I might add that the technical terms associated with filmmaking, terms the books delight in defining for their readers, form another manifestation of this status, and one transferrable to the reader.)

Such self-expression, however, requires frequent reality checks in the form of direct contact with one's constituents. In *The Moving Picture Girls* (1914), Russ takes Alice and Ruth to see one of their films in a crowded theater. They find the process both alienating and thrilling: "'I'd hardly know myself,' whispered Alice," to which Ruth responds, "'Oh! Isn't it queer to see yourself, and hear yourself criticized?'" (168). This experience is a revelation that will make the girls better actresses (171). Boy protagonists, too, need to spend time with the audience or its representatives; we see the Moving Picture Boys learning from their photographer mentor, Mr. Hadley, about the importance of attending commercial showings of their own films so that they can "listen to what the people are saying around you" (Appleton, *Boys,* 103):

> [Mr. Hadley] had said he wanted to know whether his plan of supplying daily to the public pictures of the happenings in New York and vicinity would be a success. Now was the chance to find out.
>
> In the half-darkened theatre the murmur of voices could be heard on all sides. . . . There was no doubt that Mr. Hadley had "made good."
>
> "Yes, I guess we're safe in going ahead," decided Mr. Hadley. (*Boys,* 104–6)

The public has spoken. Without their approval, the individual vision of the photographer is meaningless, as the most artistically brilliant production has no value unless the public wants it.[7] In later volumes, Blake and Joe will interview individual audience members if the collective audience is not available, as when a flood has closed down a theater that is showing their work: "'Say, they was the bulliest pictures ever I see, and I go to all the shows when I can!' he [a small boy loitering in front of the theater] cried with enthusiasm. 'They was certainly some pictures, believe me! I would like to have been there myself, only not too close,' he added with caution" (Appleton, *Flood,* 159).[8] Moreover, before he solicits this response, Blake makes clear that audience appeal and artistic merit are synonymous; the difference between what "[takes] with the audience" and "the very best sort of moving pictures" (158) is merely semantic.

Similarly, the Motion Picture Chums attend an exhibition at their new Wonderland venue, where a ventriloquist in the audience undertakes to provide sound effects appropriate to the story, and are thrilled

> [by the] furious and genuine applause. A delighted and excited old man jumped up on his chair and waved his hat, shouting:

"Three cheers for the best show on earth!"

"That was just famous."

"Must be one of those new speaking pictures."

"Oh, we must get all the folks to come to this delightful show!"

Pep's heart beat proudly as the audience filed out and he overheard this encouraging praise. (Appleton, *Seaside*, 134–35)

The boys do not measure the "tone and novelty [thus given] to the whole entertainment" (136) on some abstract scale of artistic achievement, but rather in terms of the 1,150 seats sold, as is clear from the sight of Randy "hugging his tin cash box under his arm with great complacency [as he says,] 'It couldn't have been better'" (139). In the commercial plebiscite of the box office, every admission price paid is a vote of confidence from the public, and that the Chums have outdrawn their rivals at the National Theater (who "gave away a lot of tickets" instead of selling them, padding their market unfairly, and whose seating arrangements and floors broke down because of excessive haste on the rivals' part) proves the good sense of the audience, which will award the top box-office take to what is genuinely superior.

This equation between audience appeal and general worth also emphasizes that the protagonists themselves are now commodities, which becomes obvious when audience members—a young woman in *The Moving Picture Girls*, the little boy in *The Moving Picture Boys and the Flood*—express in the protagonists' hearing the desire to meet them (Hope, *Girls*, 169; Appleton, *Flood*, 159). Inness has also noted the commodification effect present in series fiction addressed to girls, but what she emphasizes is the way that the brand name, in this instance "Ruth Fielding," comes to stand for "the unique aura" in the Benjaminian sense (176). In fact, however, more is at stake in these texts than the circulation of a brand name in lieu of the living presence of the actor or documentarian. The series books represent film performance for young women as the source of power through a deliberate commodification of the self. Thus Alice and Ruth will become better actresses through beholding their performances on screen; moreover, this distancing from the self promises to result in important information of use to the actresses themselves. For instance, in *The Moving Picture Girls in War Plays* (1916), the DeVere girls urge a young amnesiac friend to act because "a motion picture story [could] be written [around her performance]. 'It could be called "Who Is Estelle Brown?"'" Alice said, 'and it could be a serial. You could pose in it, Estelle, and make a lot of money. And not only that, but you'd find out who your relatives were, I'm sure'" (187). In short, while film enables boys to create heroic personas for themselves and to achieve success by correctly gauging audience desire, what it promises actresses is improved states of self-knowledge, coupled with the ability to circulate as money-attracting commodities.

The commodification that arises from acting's alienation of personality has several effects. The first and possibly most pervasive is, as with screenwriting, that it proves difficult to define acting as labor, given the value placed on filming actual emotion. If the heroines are genuinely terrified or distressed, can they be acting as they "register" their emotions for the camera? The possibility that the protagonists are simply being themselves tends, in turn, to elide the distinction between actor and audience. If acting merely requires feeling, then perhaps the audience is "acting" when it is moved by the film—and certainly anyone should be capable of becoming an actor, since we can all experience pity, terror, excitement, and the other emotions that form the stuff from which films are made. This point underscores the democratic nature of film.

Yet the books are loath to suggest that the young heroines are not hard working. As Maud Stanton observes of her own and her sister's career:

> We work very hard during the rehearsals and often I have become so weary that I feared I would drop to the ground in sheer exhaustion. Flo did faint, once or twice, during our first engagement with the Pictograph Company; but we find our present employers more considerate, and we have gained more importance than we had in the beginning. (Baum, 57)

The mystery plots often resolve this unhappy conflict between labor and genuineness through a slippage between the emotion that the film medium requires (which should be genuine to be exciting or moving) and the emotion that natural disasters or the machinations of business rivals generate. While the narratives often describe actors who are good at their work and therefore able to fabricate emotion competently, there are almost always plotlines designed to elicit genuine affect from the characters; the actors or other principals experience fear, horror, or dismay that cameras can then photograph to create an authentic record of emotion, and thus a more moving motion picture.

Acting vs. Accident, Fact vs. Fiction

Latent in this slippage between artifice and labor on the one hand, and sincerity and affect on the other, is the possibility of capturing an involuntary performance. One of the earliest of the series books explores this issue in the greatest detail. *Aunt Jane's Nieces Out West* examines two situations in which filmmakers essentially extort or "steal" a performance on camera, with two rather different outcomes, because the first theft involves amateurs and the second a professional actress. Beth and Patsy find themselves walking along a California street when a workman urges them back because a wall is about to collapse. As Patsy relates, "I hardly thought we were in a safe place when the big workman cried: 'There, young ladies; that will do. Your expression was simply immense and if this doesn't turn out to be the best film of the year, I'll miss my guess! Your terror stricken features will make a regular hit, for the terror wasn't assumed, you know'" (12). Later, the producer apologizes for taking advantage of the girls by explaining that his "excuse lies in our subservience to the demands of our art. We seldom hesitate at anything which tends to give our pictures the semblance of reality" (23).

Patsy and Beth are fully alive to their adventure's social ramifications. Beth in particular worries about the invasion of privacy that such a theft of her unwitting performance represents: "Meantime . . . two very respectable girls, who are not actresses, will be exhibited before the critical eyes of millions of stupid workmen, reformed drunkards, sad-faced women and wiggling children—not in dignified attitudes, mind you, but scurrying from what they supposed was an imminent danger" (16). Clearly, Beth fears that her image and Patsy's will begin to circulate among grubby workers; uncompensated labor is not an issue for these wealthy young women, but the girls acquire no authority from their performances because at this point they are outside the economic structure of the film industry. Significantly, Patsy counters Beth's complaints with an economic justification that prefigures their later involvement in film as exhibitors: "When I remember how quick-witted and alert that manager was, to catch us unawares and so add to the value of his picture, I can quite forgive the fellow his audacity" (16). That the narrative ultimately maneuvers them into a managerial role as potential large investors in a quasi-charitable theater-building enterprise may be due in part to a wish to recoup their early unpaid involvement in filmmaking. In their capacity as capitalists, moreover, Patsy and Beth are able to maintain their distance from the celebrity that destroys privacy.[9]

In contrast, their friend and colleague Maud is a professional actress. When she rescues a young man from genuine death by drowning, a rival film company photographs the event, which places her in an awkward position with the manager of the Continental Film Company, her employer. The manager complains, "I was going by one of the theatres when I noticed a placard that read: 'Sensational Film by Maud Stanton, the Queen of Motion Picture Actresses, entitled "A Gallant Rescue!" First run to-night.' I went in and saw the picture—with my own eyes!—and I saw Maud Stanton in a sea scene, rescuing a man who was drowning" (101). Only when the young man vouches for the necessity of his rescue and reveals that he is the *owner* of the Continental do Maud's employers absolve her of deliberate treachery. Crucially, Maud's performance here is doubly valuable not only because it is "genuine" but also because it is identifiable with her; she is already a commodity in circulation with film audiences, "our bright, particular star, you know . . . the public would resent it if she didn't appear" (49).

The aleatory form of filmmaking on display in these two adventures, in which narratives take shape around actual events, which the gods of misfortune seem to serve up to filmmakers in abundance, is highly characteristic of all series books dealing with film. We see this mode of story construction not only in *Aunt Jane's Nieces Out West* but even in the later *Ruth Fielding Clearing Her Name, or The Rivals of Hollywood* (1929), where a storm that might otherwise impose some hardship on the actors and crew arrives in time to be photographed, saving the company from having to go to the expense of fabricating one. In truth, feature filmmaking had long since abandoned this mode of story construction by the early teens, if it had ever much exploited it. Aunt Jane herself is a scenario reader for the Continental Company, which hardly seems a necessary line of work in a world where camera operators could successfully patrol the streets and beaches of California for naturally occurring stories.

But that the "real event" that cries out for recording and incorporation in an otherwise planned narrative remains a staple of these books until the mid-1930s suggests more than the persistence of a cliché. Motion picture series fictions are obsessed with the relationship between fact and fiction within the filmic text, not only in the frequent blurring of the two by enterprising fiction filmmakers but also in the mystery plots themselves, which, as we have seen, frequently find resolution through recourse to film as a kind of truth-telling medium. In Alger's novels, too, the hypocrisy manifest in claiming to be better than one really is characterizes the criminal, but, while Alger's villains typically pretend to wealth not their own, the villains of the motion picture series pretend to talent or achievement. These attempted thefts, of course, are invariably foiled, a circumstance that seems to imply (however implausibly) that something inherent in film itself spurns dishonesty.

Film's role as a recorder of reality has its problematic side, for it is also the means of misappropriating sensibility and experience, as the adventures of Beth, Patsy, and Maud demonstrate. That sensibility (or affect) and experience are so vulnerable warns child readers to control their emotions and themselves, since the narratives indicate that without a watchful eye one may lose these crucial aspects of the personality and suffer a corresponding loss of status, or fail to rise according to one's merits. The series books thus participate in a larger effort (illustrated, for instance, in some of the film-appreciation materials described in chapter 5) designed to arm the child audience with the means of self-mastery and to downplay the role of emotion.

Stolen performances are by no means limited to *Aunt Jane's Nieces Out West;* for example, *The Moving Picture Girls at Rocky Ranch* relates the story of an entire company's performance being "scooped" by a rival concern when a second, unauthorized cameraman follows Mr. Pertell's troupe to the oil fields of Pennsylvania (77).[10] The ease with which the canny may induce or steal performances clearly adds to the difficulty of defining acting as labor (recall Ruth Fielding's ruse in the case of Miss Calvin's initially disappointing performance). That performances can and perhaps should be stolen in the name of truth suggests that while the self ultimately has something proprietary and genuine about it, one can alienate sensibility from its owner and sell it willy-nilly, leaving the viewer unable to tell the difference on the screen. All these adventures appear linked, moreover, through their valorization of "genuine" sensibility as superior to what good acting alone might produce, which troubles both definitions of acting as labor and film's pretensions to being an artistic medium.

If filmmaking consists primarily of finding good performers and locations, and of being on the spot when something interesting happens, so that as a mode of storytelling it relies on *objets trouvés*, perhaps conventional notions of originality and creativity in works of art do not govern it. That the ease with which others may misappropriate one's work forms the basis of many of the mystery plots suggests the medium's artistic fragility, even if these misappropriations are never allowed to stand. For example, in *Ruth Fielding in the Great Northwest* (1921), Ruth discovers an abused Osage circus performer, Wonota, and translates her into a film star. Throughout the novel, both film and actress are at risk from Dakota Joe Fenbrook, who wants compensation for the value Ruth has added to his erstwhile employee, and who is willing to abduct Wonota if necessary. Ruth's creative contribution is not only the scenario but also the proper appreciation of Wonota's value as an actress and as a human being, and the realization that casting an Indian in her next film would constitute a novelty.

That Ruth is more a discoverer of worth than an originator serves to define the kinds of threats she receives from rivals, which tend typically to the theft of plans or ideas. In *Ruth Fielding Clearing Her Name*, Ruth's actresses and locations are at risk from a spiteful admirer turned rival. Mrs. Craven-Spitz starts her own film company when she discovers that Ruth won't employ her as an actress. She then operates as spoiler, hoping to deprive Ruth of cast members and California beauty spots needed for Ruth's new movie. Ruth retaliates by developing other actresses and turfing Mrs. Craven-Spitz out of areas for which Ruth has the only filming permit.[11] While the crew is in the far north during another adventure, Ruth's finished film footage briefly falls into the hands of larcenous competitors willing to deprive her company of the ice boat needed to take them to Alaska. As the narrator explains:

> Needless to say, Mr. Hammond [Ruth's producer] had many rivals in
> the motion picture industry, and not all the rivalry was amicable. As
> the fame of the Alectrion Corporation increased, in no small part due to
> Ruth's successful connection with that firm, Mr. Hammond's competi-
> tors in the field had become more bitter. There had been a *sub rosa*
> warfare going on for some time past, and for that reason Ruth would
> scarcely have been surprised at any unusual happening. (*Far North,* 36)

Simple thefts of the finished product are rarer with Ruth than with the Moving Picture Girls or Boys, however, because Ruth is more personally involved in her films as she comes to assume virtually every production role. Ruth has established herself as a brand, as her interior monologue at the time of her company's founding suggests:

> The Fielding Film Company! Plain, direct, and with the necessary advertising "punch." She meant to make the name of Ruth Fielding on the screen an important asset of the new organization. It would indicate, too (Ruth did not lack shrewdness), that the producing company was governed by her, and by no other. All publicity must naturally advertise Ruth Fielding when the corporation was thus named. (*Treasure*, 59)

As it turns out, the effect of this extreme self-commodification is to make Ruth more vulnerable to attack on a variety of fronts because of her greater personal investment in her work.

Productivity at Risk

The greatest threat to Ruth's success as an artist, consequently, comes from more intimate assaults on her sensibility. If locations and actresses are readily replaceable, one's original ideas are not. Predictably, then, even Ruth's scenario ideas are not safe; for instance, her script is stolen in *Ruth Fielding and Her Great Scenario* (1927), in which Ruth competes in a scenario contest "promoted by a large moving picture concern, a leading magazine, and a publishing house" (22) for a $50,000 prize that will, if she wins it, secure for her an international reputation and a "surplus fund" to be devoted to the "bring[ing] out [of] a new star" (94).[12] The stakes are higher, however, than choosing locations or discovering and developing new actresses. As Ruth says, "Pretty faces can always be found—ideas are more difficult to discover" (6). Her rival Dean Hollister, however, discovers Ruth's ideas quite literally when he first picks up discarded pages from Ruth's notebook as she composes the scenario, and then steals the scenario itself. Ruth at first doesn't realize how something intangible, such as her ideas, can be at risk, but a colleague tells her that her rival

> "can tell a good scenario when he sees it and he may have designs on some of your ideas."
> "You actually believe that—"
> "Yes, that if Hollister could steal an idea, he would."
> "But what would he gain? Stealing an idea isn't always profitable."
> "Hollister knows how to dispose of it once he gets his hand on valuable material." (87)

The novel implies that the usefulness of this material is only apparent to a man with some experience in the film business, which Hollister has. Ruth believes that her great contribution to cinematic art is not the plot but the "storm scene" she has been at work on, which will, among other innovations, incorporate new approaches to lighting. To understand the importance of these innovations, one must have technical expertise: "Helen, suppose some one should find the [storm] scene—someone that understands the technical language of a scenario. And just suppose that the person should decide to use the idea as his own!" (97–98). Even to be a villain in these narratives requires knowledge and the ability to discriminate between what will appeal to an audience and what will not; stealing a bad idea will not be profitable. Given the competitive nature of the business, Ruth's only protection is to be the first to the market, even with her own wares, since "if he gets ahead of me, every one will say I stole my ideas from him" (102). Thus the public perception of personality and skill potentially outweighs private truths, an idea that functions as a major source of anxiety in many series narratives.

Ruth's description of her storm scene does not, in fact, indicate any particular technical expertise, but rather a command of cliché. When Helen wonders what makes Ruth's storm scene different from its potential competitors, Ruth elaborates:

> But this is so unusual, Helen! The rascal has the child and is running away. He loses his hat in the wind. He drops the child to go after the hat. The hat rolls along the road close to the old church—the church where the child's grandfather had preached and where he was shot by the rascal. At that moment the sky darkens, the wind increases, the church steeple sways, the rascal starts to run—and the steeple comes down, burying him in the wreckage. The same wind carries the child along the road to the milk house where the girl heroine is hiding. Can't you see how Hollister might make use of all this? It's the very turning point of my scenario. (102–3)

This description confirms the difficulty of adequately defining filmmaking as labor; Ruth produces this scenario while on vacation by a lake. She represents her work as technically original, and in that sense only available to the cognoscenti. What we find, however, is an artifact that is not particularly personal and that is highly vulnerable to theft or misappropriation. At issue here is not technique but hoary sentiment, a sentiment that is all-pervasive and that tends to erase the boundaries between various roles, such as active and passive heroine or protagonist and antagonist. Not only is Ruth's idiom in this scenario (and elsewhere) melodrama, but she is the victim within the melodrama of the scenario's theft and recovery. What surprises us in this account is that this scenario appeals not only to Ruth but also to her rival, who is a hardened criminal. It seems unlikely that Hollister would originate such a story, but he can steal it, which, we are assured, will come to the same thing in the marketplace.

The Power of the Audience

Here we return to the question of whose taste is creating films in the world of one-size-fits-all authorship. While *Ruth Fielding and Her Great Scenario* assures us that Ruth's success will result from novelty and technical mastery, it is clear that, in spite of Ruth's claim that her idea is "so unusual," novelty is less at issue than is an understanding of what audiences have favored in the past. As always in series fiction, success comes from providing viewers what they crave. For example, in *Aunt Jane's Nieces Out West*, Aunt Jane notes that "the film makers are recognizing the value of the children's patronage and are trying to find subjects that especially appeal to them" (127), while in *The Moving Picture Girls under the Palms* (1914), the narrator explains, of some slapstick business, that "of course this was the farcical element, but managers have found that this is much needed in plays, and though many of them would prefer to eliminate the 'horse-play' the audiences seem to demand it, and managers are prone to cater to the tastes of their audiences when they find it pays" (122). In girls' series books as in boys', the audience is ultimately in charge of the screen fare served up. Outside the world of the series book, as well, the rhetoric of the plebiscite at the box office was an extremely important way in which the film industry manifested its concern with the popular will and demonstrated its accessibility both as an art form and as an industry. Indeed, we have here a crucial point of connection with the other case studies that this book examines: the rhetoric of the Palmer Photoplay Corporation, of religious commentary on and interaction with Hollywood, and of the character education and film appreciation movements all stressed the public's ability to "vote with its pocketbook" and its ability (or responsibility) to discriminate between good screen entertainment and bad.

If film is at times an emanation of the audience, so too may be the filmmaker. After Mrs. Craven-Spitz's attempts to damage her reputation, Ruth worries that her latest film, although good, will go down poorly, but "I won't know until after this last picture has been released. If 'The Pride of Ardale' is well received, then I'll know that my name has really been cleared" (Emerson, *Name,* 202), thus making the film's success a personal vindication; what Ruth worries about here is injury to the brand name noted by Inness. Absolute originality of both concept and personality seems to become less important toward the end of the series; Ruth graciously shares half of her unique South Sea island location in *Ruth Fielding and Her Greatest Triumph,* an act that would seem to promote the creation of similar films by rival film companies. Ruth's films will succeed in the marketplace, however, because they are hers.

Moreover, in a different example of self-commodification, Ruth seems over the course of the series to be promoting the duplication of her persona among young actresses. For example, Tom (now Ruth's husband) comments of an actress whom Ruth is developing:

> "It's obvious she's patterning [herself] after you. . . . Any one with half an eye can see she worships the ground you walk on."
>
> "She has taken a fancy to me," Ruth admitted. "She tries to imitate me in so many ways. I don't mind; in fact, I'm a little flattered, but it is a responsibility." (Emerson, *Name,* 25)

Likewise, the young heroine of Ruth's film *The Rose of Island Crag* is introduced to Ruth both as the materialization of one of her own fictional creations and as someone apparently compulsively modeling her dress on Ruth's, even, it erroneously appears, to the point of stealing a lace collar belonging to her idol (Emerson, *Triumph,* 20). Initially within the series, duplication signaled corrupt business practices; by the end, Ruth Fielding is herself a franchise to be reproduced by her actresses and admirers. The duplication is necessary to make the point that Ruth's business practices, ideas, and even persona are successful and desirable from the audience's point of view, a notion latent in the earlier novels' discussions of theft but finally ratified in the later books' treatment of emulation as forgivable when it is the tribute of a fan or subordinate rather than the machination of a business rival.

If the Alger narrative sought to provoke a certain longing on the part of the reader to emulate the success of its heroes, what are we to make of the embedded emulation of the fan or jealous business rival in the

Ruth Fielding series? Moon argues of the Alger story that it offers a "modest-demand, modest-reward ethos" (89), an approach obviously troubled by the spectacular social rise possible in the motion picture industry for characters such as Ruth Fielding. Ruth's rewards are huge, and the emotional demands, in particular, are commensurate with this scale. The ethos of self-cultivation results, when successful, in a product so alluring that everyone will wish to possess (or be) it. Emulation, in effect, is another manifestation of the "vote" at the box office, a ratification of the public's ability to discriminate among more or less successful film workers. This very alienation of the commodified self from its presumed single possessor leaves Ruth open to assault.

Repression and Rise

If Ruth's livelihood depends on being able to manipulate sentiment in cast and audience, self-cultivation and self-discipline are the tools that allow her to control the new emotional demands arising from the development of her star persona. Ironically, as a purveyor of affect, Ruth gets ahead precisely because she represses her own emotions. Even at the beginning of the series, when she is an orphan and not yet a screenwriter, she fights back tears of loneliness when leaving her friends for her new life. Later, as a famous writer, actress, and producer, Ruth continues to succeed by maintaining command over herself, thus modeling the kind of self-mastery that these series inevitably recommend. If personal power results from one's affiliation with film, it is essential that this affiliation go hand in hand with self-control, repression of affect, and the acquisition of technical skills, all of which justify self-commodification by fighting off others' attempts to commodify and exploit one. For commodification by others can be dangerous. So successful is Ruth's self-cultivation that the texts attribute various assaults on every aspect of her property and persona to malevolent desires to emulate or possess, even when facts eventually prove the contrary. When, in an eerie prevision of the Lindbergh case, Ruth's baby falls prey to a kidnapper in a 1931 installment, one of Ruth's friends suggests that her fame has made her vulnerable to attack: "Sometimes it isn't enjoyable after all to be so gifted—to be so famous—to be in the limelight and get so much publicity all the time" (Emerson, *Baby June*, 182).

Here, Ruth's self-control and ability to see real events as the stuff of film scenarios ultimately enable her to foil her baby's insane kidnapper.

After the madman has abducted her child, Ruth rescues the baby by pretending to be a nurse casually hired to take care of her:

> A moment more and she had Baby June in her arms. But even yet she must act. She must not cuddle her child too closely, yet with all her being Ruth longed to press Baby June to her heart. She must not kiss her. She must still pretend indifference. And in a voice that had endeared her to the talking-picture directors Ruth said:
> "A nice little baby."
> For a split second the doctor fairly glared at her. His eyes smouldered with hidden fires. Had she over-acted? Had she overplayed her part? But her calm, careless air had its effect; disarming the fellow. (Emerson, *Baby June*, 197)

Inness reads this scene as evidence that series fiction prepared young women "to be active agents rather than passive spectators" (173). It is also possible, however, to view this moment as endorsing self-control and ratifying the commercial value of affect. While Ruth is not performing for money here but rather for the safety of her family, she clearly draws on her commercial past to accomplish the rescue. But that she is still vulnerable to assaults on her sensibility and emotions in a narrative that does not contain an interpolated film is striking. Even the omnicompetent Ruth, it would appear, is the victim of an extorted performance; she is obliged to enact the role of nurse when she would rather play mother. Yet Ruth's suppression of the "appropriate" emotions secures the safety of her baby. It may, moreover, work to suggest that the role of working woman is compatible with that of good mother, since Ruth's skill in the public sphere proves crucial in the private sphere. Although Ruth leaves the professional realm at the conclusion of the series, these books never criticize her for remaining devoted to work even after she bears a child.

Something still more fundamental is at stake in this kidnapping narrative, however. Not only does Ruth's professional prowess come in for praise while demonstrating its utility in the private sphere, but the combination of powerlessness (what can an unprotected woman do to save the life of her child when being menaced by a madman?) and competence (she can act, in two senses) also suggests an interesting point of contact between Ruth as star figure in this series and the actual stars fostered by the film industry. Lary May argues that film stars diffused the potential ill will built up owing to their positions as "idols of consumption" (in Lowenthal's phrase) by being remarkably powerless in other areas, such as politics: "the movie folk could be universally loved because they were not socially powerful: they were purely a status group" (197). In fact, Ruth Fielding, like Mary Pickford, is a more complex figure than this insight would suggest because she wields economic authority as a studio head. And, indeed, some of the assaults she endures (such as her own abduction in *Ruth Fielding in Talking Pictures*) are a result of her exercise of that authority, as when, for example, Ruth decides not to hire the man who eventually kidnaps her. Nonetheless, her star persona also seems to leave Ruth open to attack. In *Ruth Fielding and Baby June*, Ruth suffers because of decisions any (wealthy) mother might make in choosing servants (the madman is the estranged husband of her child's nurse). That Ruth's professionalism receives its reward in that she retains the mastery she has so laboriously acquired suggests that she cannot be taken advantage of; even when circumstances extort a performance from her, Ruth can turn it to good effect because of her training in the suppression of her own feeling and the manipulation of feeling in others. Through her "professional" performance, she in effect rewrites mad Dr. Grenade's script.

This way of reading this adventure has two ramifications. The first is that stars must pay for their success, through the emotional suffering they endure as a result of admiration. While the novel ultimately suggests that Grenade has other things on his mind when he steals the baby, Ruth's associates assume that she is the target because she and everything about her are known and desirable. (And, we might add, even the baby has already become a commodity through her participation in a "beautiful baby" contest, which establishes her desirability by awarding her the prize for "best in show" and thus makes her an object worth stealing.) If Ruth is vulnerable because she expertly manipulates affect in both employees and audience (thus calling forth strong emotions both enjoyable and frightening), further manipulation of affect is the cure for the problems that ensue. The second ramification is that Ruth's perfect emotional self-command, something that the new workplace will demand of women whatever their profession, is essential to the project of controlling affect in others. It is the very mark of the professional and the only way that one can use emotion safely. *Ruth Fielding and*

Baby June insists that Ruth feels but that her feelings will not get in the way of her aims, which gives a new gloss to the oft-repeated slogan of the series, "To think was to act with Ruth Fielding."

The series fictions under discussion thus engage in an extensive discussion of the self's relation to intellectual property, to sensibility, to genuine affect versus fabricated emotion. The reader discovers that filmmaking sees no hard and fast separations between the fictional and the real, and that the sine qua non of a performance is its authenticity—a point, as I have noted earlier, that tends to elide the boundaries between fictional filmmaker and real audience member. If not all interpolated performances are elicited through provoking real emotion from actors, then the novels typically provide their protagonists with experiences separate from the embedded film narratives that call for the same emotions required by those embedded narratives. The reader gets the genuine affect one way or the other and learns to accept no substitute from actors. Moreover, if affect is always preferable to skill within performance, then little separates the reader as a potential star from any character in the novels. Of course, the protagonists of the novels remain superior through their grit, their good fortune in being in the right place at the right time, and their detective ability.

While filmmaking comes across as a highly skilled, technical profession, the books base their appeal on the appearance of being able to give away the mysteries, thus reducing the gap between reader and adept and theoretically democratizing the discourse. The advertising of the Moving Picture Boys series notes that "moving pictures and photo plays are famous the world over, and in this line of books the reader is given a full description of how the films are made—the scenes of little dramas, indoors and out, trick pictures to satisfy the curious" (quoted in Johnson, *Pseudonyms,* 20). But in fact the books are neither comprehensive nor completely up to date, and even the earliest titles were reprinted in many forms into the late 1920s, when the information they offered was creaky with age.

As I noted earlier, Fuller cites these series as an aspect of fan activity and the fruit of savvy marketing of "the hobbyist aspects of film exhibition to young tinkerers" (125). Who can deny that the project of making teenagers giants of the film industry suggests a peculiarly accessible career, especially if we are willing to stipulate that the books serve as a useful introduction to the technology of filmmaking or the structure of the industry? Yet most of the young heroes are larger than life in their achievements if not in their native ability, as Carol Billman observes when she comments that "given Ruth [Fielding]'s spectacular mobility (both geographical and social) and her access to the risks, mysteries, and opportunities of the wide world, there is ultimately no way the character could be read by girls and young women as one of them" (73). These characters create for themselves (sometimes with smoke and mirrors) personas that exist somewhere between the heroic and high mimetic modes, to borrow Northrop Frye's categories, and as such are beyond the reach of the average child.

But perhaps this last observation explains exactly why the books proved so popular for so long, and why, given the large number of authors at work under the several pseudonyms and the extended timespan from first volume to last, the anxieties that the various series explore seem remarkably consistent. The books are at their most powerful not in encouraging children to enter the film industry, or in representing the industry as the cure to commercial ills, but in suggesting that even the successful suffer assaults on their sensibility and affect. In fact, the characters on the whole are successful not because they suffer, or in spite of their suffering, but because they can turn their own and other people's affect into power, money, and status. Indeed, through their demonstration of "how" motion pictures work, the books seek to reveal not specific technical knowledge, but rather the alchemical secret of turning feeling into money. The outsider, the child reader, becomes the insider not through the acquisition of facts but through the appreciation of the salability of emotion and the concurrent understanding that, for the vendor of emotion (or the consumer of film), self-control is desirable.

Conclusion

Ann Cvetkovich cautions that the expression of feeling is not inevitably liberating for a given social group (women, for example), since "rather than leading to social change, the expression of feeling can become an end in itself or an individualist solution to systemic problems" (1). The suppression of feeling may or may not be liberating in the same fashion. What we encounter in series fiction is the manipulation of affect partly in the guise of a solution to a social problem (as when film documents the abuses of evil capitalists), but more characteristically as the means for individual social rise. Almost inevitably concomitant with that elevation is the discovery that protagonists must control their personal feeling even as these budding film artists or exhibitors become more adept at shaping the feelings of others. If the suppression of feeling results in material advantages, it also demands a self-discipline more severe than most employers would ask of those who sell their labor, as opposed to their "selves." Series fictions expose the price of self-commodification but argue that it is well worth paying for the social power it brings to the individual. As will become apparent in the next chapter, the Palmer Photoplay Corporation materials would continue to attach the rhetoric of self-commodification to the hope of *individual* social rise. Both series fiction and the Palmer Photoplay Corporation invite the individual to retool him- or herself as a possible member of this new corporate, technological sphere through the application of "technique" to both labor and self, often through the control of emotion.

And it is through such endeavors, according to both sets of narratives, that the marginalized outsider, the ninety-eight pound weakling who labors under the disadvantage of being an orphan or a child or a girl or poor, may transform him- or herself into a powerful industry insider, master of the self and others. In this regard, the rhetoric of personal power constructed within series fiction and correspondence-school come-ons alike fits seamlessly into an American myth of social rise that predates and outreaches Hollywood, a myth extending from Alger to Charles Atlas and beyond. Or in Daniel Boorstin's words, "The film-star legend of the accidentally discovered soda-fountain girl who was quickly elevated to stardom soon took its place alongside the log-cabin-to-White-House legend as a leitmotif of American democratic folklore" (156–57).

Finally, the ratification of the elevation of the meritorious adolescent by the filmgoing electorate is itself a confirmation of the successful operation of democracy. Fans and protagonists are locked in a mutual embrace that endorses the good judgment of the crowd in preferring these plucky children to their rivals, even as it suggests that those in charge of purveying entertainment for the masses have the best interests of the audience at heart. Indeed, the purity of the heroes and heroines guarantees the good conduct of the film industry, even as their sacrifices entitle them to lead it. In their suppression of inappropriate emotion and their embrace of professional standards, these children prefigure the concerns of the film appreciation movement discussed in chapter 5, a movement whose pupils will be taught to resist the blandishments of the film industry through a similar ethos of self-discipline and incorporation of standards. In a perhaps unexpected reiteration of the slippage between the real and the fictional, however, we encounter, in the case of the film appreciation movement, an interesting inversion of the process by which feeling is controlled. If series fiction protagonists generate audience affect by substituting the real event for the fictional composition, film appreciators attempted to reduce film's ability to create excessive feeling by reminding their students of the "real" world behind Hollywood's fictional constructions.

CHAPTER THREE

Fashioning the Self to Fashion the Film
The Case of the Palmer Photoplay Corporation

> Great poets are often proverbially ignorant of life. What they know
> has come by observation of themselves: they have found within
> them one highly delicate and sensitive specimen of human nature,
> on which the laws of emotion are written in large characters, such as
> can be read off without much study.
> —John Stuart Mill, "What Is Poetry?"

In *Babbitt* (1922), Sinclair Lewis provides his readers with the text of an advertisement for a correspondence course in public speaking that has caught young Ted Babbitt's eye, and which Ted hopes to convince his father to allow him to take in place of the more conventional college degree. The language of this advertisement, with its emphasis on "personality" and its enthusiasm for power over other people, suggests that education was developing a new instrumentality in the 1920s. As *Babbitt's* narrator notes, "above the picture [of a young man holding an audience of middle-aged men rapt] was an inspiring educational symbol—no antiquated lamp or torch or owl of Minerva, but a row of dollar signs" (65).

While the courses that attract Ted focus on traditional subjects (from oratory to boxing), correspondence schools sold expertise in newer fields as well. During the 1910s and 1920s, a number of such schools promised to prepare men and women for jobs in the film industry, especially as screenwriters. The Palmer Photoplay Corporation in Los Angeles, founded in 1918, was the largest and most successful of these outfits, which flourished in large and small cities across the nation. Cynically viewed, these schools might appear only to be in the business of separating customers from their money, since after the freelance market collapsed in the mid-1910s, the likelihood of a private individual's placing a screenplay with a studio was small. The thriving nature of Palmer Photoplay's business as late as the mid-1920s (when it began to shift its attention to instruction in short story writing and general self-expression) suggests that achieving material success in Hollywood may not have been the sole motive for subscribers, or indeed for the school's management.

Palmer Photoplay was the brainchild of Frederick Palmer, a thirty-seven-year-old self-described Cornell University graduate[1] who had begun his career as a newspaper reporter, an editor, and an actor. Palmer Photoplay located itself at the crossroads of contradictory rhetorics about the nature and destiny of the American film industry in the early 1920s. These oppositions—democratic vs. elitist, individual vs. corporate, private vs. public, and original vs. hackneyed—structure this discussion of the Palmer Photoplay Corporation between 1918 and 1925. Palmer's express aims ranged from the fostering of a sense of kinship between audience and producer to addressing defects in "personality" and general education. Beyond these overt motivations, however, lay the less frequently expressed desire, apparent also in the series books discussed in chapter 2, to create a "better" audience, more engaged, more informed, more invested in Hollywood—in short, an audience knowledgeable about and thus sympathetic to the industry and the medium. In a slippage characteristic of the 1910s and 1920s, it is not things or products being mass-produced but people. If Henry Ford asserted that he manufactured men (through the Americanization of his workers) rather than motor cars, Palmer Photoplay Corporation manufactured superior audience members rather than screenwriters *tout court*.

To this end, the school worked to promote a notion of self-expression simultaneously standardized and individual. The present chapter will examine Palmer rhetoric consisting of advertisements, direct-mail solicitations, textbook prose, and corporate self-description, as well as other discussions of Palmer and of freelance scenario writing generally, to consider how the twin goals of standardization and individuality affected the school's dealings with its students and helped to shape Palmer's presentation of filmmaking as a democratic field. The related questions of Palmer's clientele, mode of instruction, and ambitions help to clarify the school's position on how the outsider might most readily become a Hollywood insider. Here Palmer reprises the strategies promoted by the series fiction inasmuch as the corporation's rhetoric suggests that the manipulation of affect is the means to success, a process that, again, requires the aspirant to engage in a commodification of the self. Palmer's sales strategy suggests that while standardization was the key to turning students into successful screenwriters, the apparent promotion of individuality was the key to recruitment of students in the first place. The divide between standardization and individuality fostered by advertising and by the culture of personality intersects in complex ways with the film industry's own discourse of standardization and differentiation, highlighting ways the cathexis between consumer and good appears to require or generate the desire for participation.

The standardization of Palmer students began with a mandatory questionnaire, developed by a screenwriter and a psychologist and designed to divine talent for photoplaywriting (*Essentials,* 9).[2] A number of approaches potentially worked toward a kind of Taylorization of self-expression: student work was to be presented to potential buyers in a uniform format; the school's stock in trade was instruction in "technique"; and students were advised as a group to avoid certain topics (costume plays, war and Bible pictures) and stylistic fillips (excessive reliance on melodrama and coincidence, for example). At the same time, however, the school preached the gospel of individual success through self-fashioning, the importance of originality, the parlaying of personal experience into audience affect, and the evils of the Eastern publishing establishment and of industry reliance on adaptation. As already discussed, the notion of individual social rise is itself potentially Taylorist in that it reflects the elevation of the successful worker foreseen by Taylor in his bonus system providing incentives to the most productive workers. Ironically, submission to Palmer's corporate enterprise was presented as a superior means of individuation and as an escape from unsatisfying work in other fields.

These competing rhetorics do not arise merely from Palmer's advertising, nor do they result simply from the quixotic mission of preparing students for a freelance screenplay market that had largely disappeared by the time the school was founded. Rather, the rhetorical contradictions manifested by the school's materials are likewise found in trade paper and general circulation periodical articles discussing the nature and destiny of the film industry. The "original" story, neither an adaptation nor hackneyed, was held out as the ideal to which the industry must ultimately adhere; that many commentators and industry practitioners (by no means all shills for correspondence schools) discussed the dearth of original material may have suggested to the would-be author that an opening existed for him or her in spite of indications to the contrary. Similarly, journals such as *The Film Spectator* called for instruction in the arts associated with the film industry, a call that may have promoted the notion that this field was not yet overcrowded and might be colonized by the outsider prepared to learn a new skill.[3]

The place of the "pre-sold" film property, the question of the desirability or undesirability of adaptation, and concerns over the intellectual development of audiences and the place of film in American culture exercised the public in many venues during the early 1920s. As a node within the nexus of these debates over the social role of the film industry, Palmer was in Hollywood but not exactly of it. Many of its personnel had connections to studios, and its advisory council boasted such luminaries as Cecil B. DeMille, Lois Weber, and Thomas Ince. But while Palmer dabbled in film manufacture between 1922 and 1924, producing films known as Palmerplays and authored by its students, it was not a studio in the conventional understanding of that term. Rather, Palmer offered itself as a liaison between Hollywood and the American public, addressing its students primarily as would-be members of the industry. Indeed, its function as agent was arguably its most seductive, suggesting to the outsider that he or she had a representative within the film industry smoothing the path. As a structure designed to encourage outsider participation even as it controlled it, Palmer operated in a way analogous to that of the Hays Office, which used the same rhetoric of accessibility tempered by economic and narrative exigencies. Consider, for instance, a form letter from Frederick Palmer informing potential students about the nascent film production scheme: "Our entry into the producing field is designed to break down the last barrier between the new writer and the motion picture screen" (letter of 29 August 1922 to F. C. Baker). This letter suggests that the corporation could unite outsider and industry for their common benefit by explicating the secrets of authorship to interested members of the public.

The Industrial Context of Screenwriting

Further discussion of the industrial context within which Palmer Photoplay operated may help in evaluating what it claimed to be able to do for its subscribers. In a way reminiscent of juvenile series fiction, Palmer exploited an understanding of the film industry more relevant to an earlier era. Indeed, in presenting Hollywood as a comparatively open market eager to reward the freelancer and the creative outsider, Palmer capitalized on a perception of the industry as a haven for workers fleeing regimentation elsewhere even while suggesting that the way to succeed in this demi-paradise was either to submit to the acquisition of exacting technique, the means of turning personal experience into affect into money, or (as the protagonists of the series fiction also discovered) to engage in a process of self-commodification for profit.

In fact, from the late 1880s to the late 1910s, the film industry rapidly matured into an oligopoly. But what concerns us here are not institutional structures so much as the ways their development impinged upon representations of the film industry as a field open, particularly in the area of screenwriting, to interested outsiders. In moving through the arc from open endeavor to centralized business, the industry participated in the industrial change taking place throughout the American economy; film's comparatively late start at the end of the nineteenth century, however, caused it to appear to some observers as temporarily out of phase with the changes elsewhere, and consequently gave it a reputation as the antidote to the increasingly rigid organization that was shaping labor in fields as diverse as insurance and education.

Janet Staiger has thoroughly pursued the issue of the development of modern business practices within screenwriting. As she notes, the film industry had to reconcile the twin but opposed goals of standardization, which promoted efficiency, and differentiation, needed in an industry whose lifeblood in marketing was "novelty." The script was perhaps the area within film manufacturing where this tension was felt most keenly, because each script was the blueprint for the finished product. Cost control, efficient use of labor, and the ability to consolidate quality standards all followed from standardizing scripts, and such standardization, although apparent even earlier, was a noticeable feature of studio practice by 1911 ("Photoplays," 15–16).

In practice, standardizing scripts meant developing common understandings of what a photoplay was, not only among the craftspeople who made it but also among the public who watched it. Significantly, the public had the opportunity to follow these developments through the trade press and fan magazines, which codified and disseminated the "inside," technical knowledge that the film industry was developing for its practitioners (Bordwell, Staiger, Thompson, 194). Approved story-telling elements included well-defined characters, psychological rather than merely physical causality, an emphasis on action, and the privileging of an intelligible story that would control viewer attention and maximize comprehension (Bordwell, Staiger, Thompson, 180–81 and 195). The techniques by which stories were constructed were intended to be unnoticeable while one watched a film, but the amateur writer was nonetheless expected to follow them if he or she hoped to offer any work to the film industry. In this respect, acquaintance with the new streamlined literary form of the scenario (which was both like and unlike its nineteenth-century antecedent, "the well-made play") required the same kind of deliberate public appreciation that, observes Neil Harris, other industrially designed objects such as cars and household goods were receiving (187). The informed consumer, in other words, was to be a connoisseur of product design. Moreover, new industrial design in any form tended to hold at bay the decorative aspects of bygone styles in an attempt to present modern technology to the public (Harris, 187), and the scenario was no exception.

It was in taking a story idea and reworking it in "continuity" form that the process of standardization was most apparent—and that the new "technical" background was most essential, although at a level that the amateur might not be expected to follow. Story ideas, in contrast, were the area most dependent (at least in theory) on novelty, and were the place in scriptwriting where the enthusiastic but untrained audience member could perhaps make a contribution. Originality, as Staiger observes, was a major emphasis within film marketing, where it participated in a larger network of discourses about what kind of cultural enterprise filmmaking was; such marketing also included conspicuous markers of "value" ranging from the use of big-name actors to the elaborate use of technology and the manufacture of spectacle ("Photoplays," 20). Indeed, the seamlessness of film technique was another marker of cultural value, one that tended to affiliate it with middle-class entertainments such as the legitimate stage. These discourses of cultural value consequently might be said to obscure the discourses of economic efficiency, although the savvy filmgoer would be aware of these two opposed ways of looking at any given film product. Palmer's literature strongly suggests that part of the appeal of any information addressed to the would-be insider was the promise of mastery inherent in the revelation of economic and industrial secrets; in presenting such revelations, these discourses participate in the similar pleasures offered by juvenile series books or by fan literature about stars.

The tension between standardization and differentiation has given rise to extremely complicated rhetorics about the film industry. This complexity follows from the industry's desire to appeal to competing authorities by making two contradictory statements about its nature and any given instance of its product—that filmmaking is both industrial production and art (or both egalitarian and elitist). Nowhere is this tension so evident as in the gap between the studio's continuity staff and the professional writer with credentials formed outside the film industry. To summarize Staiger's findings, some production companies had begun to develop staffs of writers by the early 1910s, a practice that maintained predictable production volume, avoided the pitfalls of possible violations of copyright and the costs associated with purchasing texts from other sources, and secured the exclusive services of valuable writing talent. Even the category of studio author had two levels, that of the supplier of the story idea and that of the continuity writer proper; the latter was by reputation the more lowly and mechanical position, even while it may have been possibly more informed by "technique" than was the position of "idea man." The development of the "tame" author, Staiger notes, led to a dramatic downsizing of the freelance market by 1916 or so ("Authors," 34–35), although Palmer promoted a notion of the film industry's accessibility to the nonprofessional through the strategy of turning the amateur into a quasi-professional.

Indeed, the plausibility of Palmer's continued invitation to the public to contribute to the film industry is probably an artifact of a brief period in the early 1910s that apparently did indeed see a dearth of stories available for filming; it should be noted, though, that Palmer's claims in 1919 that this shortage was still in effect (advertisements in *Photoplay* April 1919, 98, and May 1919, 98) may have had some justification where some studios, such as Famous Players-Lasky, were concerned. Edward Azlant suggests that the 1911 Supreme Court ruling that found literary properties to be tangible goods, and consequently protected from unauthorized adaptation, resulted in the temporarily increased commissioning of story material from all sources, including the general public (104). Following the collapse of the freelance market, however, stories were typically agented (a role that Palmer offered to assume for its students), legally adapted from novels or magazines, or commissioned from the in-house staff (Staiger, "Authors," 42), although solicitations of material from outsiders could still occasionally be found in trade publications such as *Variety*.

Agents appealed to the film industry because they in effect created a provenance for the work they conveyed to it and thus frustrated claims of plagiarism. By 1925, an MPPDA publication on screenwriting addressed to the public noted that most studios no longer accepted unsolicited submissions because of the risk of litigation that such manuscripts presented ("Facts"). Studios took elaborate precautions by the late 1920s to avoid plagiarism suits, which could be quite costly (see "Unsolicited Manuscripts"; "$500,000 Sought").

While story material produced in-house had the advantages of low cost and greater reliability, it frequently did nothing for the claims of art, not because it was necessarily inferior but because it was concocted under industrial conditions that undercut its claims to being anything but a convenient manufacture. When confronted with the charge that movies were not art, the film industry's most characteristic strategy for upgrading its product was to yoke its efforts to some better established artistic enterprise, such as the stage or publishing, either through purchasing works already proven in these venues or by co-opting the labor of workers from those more respectable fields. This invocation of the authority of high culture not only permitted but demanded an invasion of Hollywood by industry outsiders. Samuel Goldwyn's Eminent Authors project was such an attempt to improve film's status (while simultaneously differentiating Goldwyn's product from that of rival Jesse Lasky's "Famous Players in Famous Plays"); Goldwyn's efforts suggest that the strangely divided rhetorics surrounding Palmer Photoplay were not solely the product of manipulation on the part of the correspondence school but also arose, to some degree, from the equally divided representations of the film industry proper. Both Palmer and Goldwyn hoped to bridge the gulf between insider and outsider status, just as both grappled with film's connection to other art forms.

The appeal to the authority of art, however, was necessarily in conflict with concurrent appeals to the authority of the public (manifested in the proto-audience research described in juvenile series fiction and in Palmer prospectuses) and to the authority of the industry itself.

Eminent Authors, for example, imported a number of "Eastern" writers to California in the name of art, but they appear to have made little contribution to filmmaking during the period of the experiment (1919–23). This well-intentioned but doomed transplant of talent highlights the very problem of cultural authority that it hoped to resolve. Mary Roberts Rinehart, probably the most successful of the imports, believed that Goldwyn's program was deliberately undermined by his continuity staff, who did not want to see Eastern imports succeed where they had not (Fine, 52–53). The failure of the Eminent Authors program was an early but significant manifestation of what film outsiders felt was a self-interested and "entrenched bureaucracy," as John Schultheiss puts it, that prevented both the progress of art and the personal development of the promising outsider, whether professional or amateur (Schultheiss, 16).

If the scenario department was hostile to established writers, its attitude toward the nonprofessional was even less welcoming. Richard Koszarski quotes, as a description of the loss of quality control that caused the collapse of the freelance market, an anonymous diatribe appearing in the February 1919 *Bookman*, "The Movies: A Colossus That Totters" (108). But the article may be more significant for its picture of the hostility awakened by the (largely female) scenario department:

> To begin with, the editorial department of a picture concern has an outer room where a clerical staff of young women opens and files incoming mail. . . . When manuscripts come in they are handed over to the reading department. This is a room where half a dozen or more young women at an average wage of ten dollars a week, without the competence of a stenographer or salesgirl, sit all day making first choice of the material the editor is to see. . . . Whether there is some vague notion on the company's part that these young ladies represent the typical motion-picture going public mind or not, I cannot say. ("Movies," 655)

While women are not the only sex condemned here, they are accused in particular of the perennial sins of scenario departments, piracy and bias: "Each young lady in her sub-editorial chair has a scenario or two of her own up her georgette sleeve; a scenario stolen, compiled, consciously or unconsciously, or 'original,' which she is determined shall have preferment over outside submitted material" (656).[4]

Piracy in particular exercised both the public and the industry since this activity heightened the instabilities latent in a Hollywood enterprise poised between craft and industry, high art (potentially) and popular entertainment, centralized business and participatory endeavor. The concern over creativity (and whose creativity) latent in the anxiety over piracy suggests that "originality" occupied a central place in any discussion of the relation between Hollywood and its audience. Was the popular audience the source of personal stories that would rejuvenate a stale medium, or was it the impediment to greater artistic success by an industry that would never develop as long as it pandered to the masses?

Palmer entered this debate by suggesting that Hollywood had a place for "trained" members of the public, professionalized outsiders who would preserve the necessary common touch even as they mastered the narrative and economic standards shaping the film industry. The standards themselves were seen as promoting democracy by suggesting that merit would win out over bureaucracy or elitist notions of art; Hollywood would welcome the talented neophyte. Yet the standards also worked to shape "personality," proving their value in creating "fascination" whether or not a screenwriting student ever succeeded in selling a story, and in promising that the acquisition of filmic skill would enhance the individual's personal power. This situation hints that the desire for elitist structures had not gone away so much as it had been redefined. Thus the Palmer rhetoric creates alliances between, for instance, "Eastern" writers and amateur screenwriters because both are shut out of success by a studio functionary, and simultaneously implies that the willing Palmer student has it all over those with more conventional training, because he or she is apprenticed to industry insiders who will impart only the technical information needed, while "Eastern" writers are crippled through excessively "highbrow" education.

The Palmer Empire

Even the manuals and agenting services that Palmer offered to its students participated in the debate about the place of other sources of artistic legitimacy in filmmaking. The instructional materials shored up weaknesses in general education and were remarkably ambitious in scope, even as they suggested that all the education a student needed to write well was the technical instruction supplied by the school. During their period of study with Palmer—typically a year although the term could be extended—subscribers received numerous books (including the *Photoplay Plot Encyclopedia*, described as "an analysis of the use in photoplays of the thirty-six dramatic situations and their subdivisions" [Palmer, *Encyclopedia*, 3]), twelve lectures, and a newsletter reviewing current releases, as well as a monthly magazine, *Photodramatist,* giving advice on scenario writing and observations on Hollywood.[5] In addition, the organization provided exercises to work through and a revision and placement service for manuscripts. Subscription rates ranged between $55 to $75 in the 1910s and $76.50 to $90 in the 1920s, depending on which plan students chose.[6] Students were entitled to send in five manuscripts to be critiqued; the placement department would undertake to sell acceptable manuscripts on commission. Moreover, between 1922 and 1924, students might hope to have their films selected for production as Palmerplays, although the chances of this happening were admittedly slim.

In contrast, many competing correspondence schools offered a single book or a subscription to a magazine in exchange for a couple of dollars. Such rivals to Palmer included the Photoplay Enterprise Association in Boonville, Indiana, which evidently consisted of one volume, sent through the mail for $1.50 and authored by Monte Katterjohn, who also offered to critique and agent scripts sent to the Association; the Home Correspondence School in Springfield, Massachusetts, which published *Writer's Monthly,* among other titles, and resembled Palmer in that it would critique and agent students' work;[7] and the Manhattan Motion Picture Institute in New York, which sponsored Florence Radinoff's *The Photoplaywright's Handy Text-Book* (1913). No program unaffiliated with a college or university offered anything as elaborate as the Palmer plan, which traded not on academic prestige but on its closeness to the film industry. In comprehensiveness, Palmer's closest competitor was probably Columbia University's extension program, where Frances Taylor Patterson and Arthur Edwin Krows taught screenwriting from the early 1920s through the late 1930s.

The elaborateness of the Palmer program no doubt assured students that they were getting the real thing; it also recalls the omnicompetence of writer-director-producer-star Ruth Fielding. Like a real-life version of this fictional figure, Frederick Palmer has commodified and diversified himself by "becoming" a company. (Is a "Palmerplay" a script produced under the corporation's aegis or one ultimately authored by Frederick Palmer as mentor/muse, or both?) Just as Ruth gives her name to her production company so that every subsequent film will advertise *her,* Palmer is not merely an individual but a national brand, as is suggested by the text of the earliest advertisement (soliciting both authors and producers) that I have located: "Scenarios that carry this stamp ['read and approved—Frederick Palmer'] are instantly recognized by producers, directors and editors as stories of unquestioned merit" (*New York Dramatic Mirror*, 21 September 1918, 457).

The Palmer Curriculum

Most screenwriting textbooks, Palmer's included, offered the same practical advice to students, typically in the form of a codification of the classical Hollywood style. Frederick Palmer's advice to his students was so clearly in line with the prevailing norms of the classical Hollywood style that historians cite his pithy summaries of what the scenario is as evidence for the rationale behind the style (Bordwell, Staiger, Thompson, 15 and 177). In less elaborate form, however, much of the Palmer information was also available in trade papers, books, and fan magazines, reflecting widespread interest in screenwriting during this period.

Among the pointers about classical technique provided to students was the notion that action is indispensable to the photoplay,[8] a dictum connected with a mantra of the advice columns, namely that things must be visible and visualizable (Sarver, 9). It is at times difficult to distinguish between two possible meanings for "visualize" in the screenwriting parlance of this period: the first, to tell an action-packed story with obvious causal links, and the second, to tell a story visually, involving narrational devices such as eyeline matches. Inasmuch as the technical vocabulary of editing and camerawork only rarely appears in advice columns, the first definition dominates the discourse, although at moments elements of the second intrude. Interestingly, however, while the "visualizable" was Palmer's stock in trade, the school's promotional literature took pains to suggest that a lack of gifts in "visualization" need not prevent ultimate success. As a school hoping to improve its students (and convince them that subscribing was worth the expense), Palmer naturally emphasized characteristics amenable to correction.

In addition to the importance of action, the correspondence schools' lectures and manuals explicated methods of developing profitable characterization and of structuring suspenseful plots. Other pieces of advice were pointers on how to prepare manuscripts and where to send them (including a regular *Photodramatist* column on the state of the market). The instructional materials also underlined the visual nature of film;

among the more arcane assignments is a prolonged exercise in "visualization" (in the first of the two senses discussed above), in which the student is to imagine a room in detail, people it, and then place these imaginary figures in conflict with each other. But this assignment may be less a way of learning how to write than an opportunity for the exercise of mastery: "Start visualizing in a passive mood.... Then resume control. Become master of your visualization, bringing logic and reason into action.... in a short time you will experience the magic of becoming a monarch of the limitless domain of your own imagination" (Palmer, *Handbook* 1918, 51).

What Palmer offered its students, in other words, was to some extent mastery of the self, a way of getting one's daydreams under control and turned to profitable use. The rhetoric of mastery tended to rely on industrial or scientific metaphors for its persuasiveness. When a Palmer advertisement asked potential students, "Have *you* the power that makes some men and women great?" (*Photodramatist* October 1922, 36), it clearly hoped to harness the general enthusiasm for industrial success that crops up in so many 1920s advertisements, to the specific desire to achieve fame through screenwriting. "Creative imagination" is the tool that will best assist students, and it is strangely general in its application: "It is the force that solves most of life's problems; that builds great dams, factories, and universities; that produces x-rays and radio; that writes masterpieces of literature." It is no accident that every example but one belongs to the world of industry, science, or corporate life. Thus the reader is assured that screenwriting is not so much an art as a form of engineering. This activity rests upon the absorption of "technique," represented as a body of knowledge separating the tyro from the adept; the course thus takes on the mystique of a kind of freemasonry, the charm of belonging to an inner circle of industrial professionals.

In some instances, technical mastery is described as a new form of linguistic competence, which may range from learning the vocabulary of filmmaking to beginning to think in a new, more visual language. As Frederick Palmer observes of his own development as a writer after achieving new understanding of photoplay construction, "not only did I immediately begin to see a new light, but I really began to think in a *new language*" (*Essentials*, 5; Palmer's italics). This language clearly represents a technical skill through which students can turn their ideas into marketable items, thereby engaging (on a smaller scale) in the same self-commodification that Palmer is modeling. New competence aside, however, students are advised to eschew the very vocabulary they pick up from the Palmer Plan: "Do not attempt to use the technical terms contained in the Glossary when preparing a synopsis for submission. Tell your story in clear, simple language" (Palmer, *Handbook* 1921, 10). This direction suggests the double-barreled appeal of the Palmer Plan as simultaneously a tool for self-construction and a practical, down-to-earth way of communicating with the masses.

Telling one's story in simple language did not necessarily rule out the use of melodrama to produce affect in others. It is important, as Martha Banta argues in *Taylored Lives,* that the literary vogue for professional and scientific narratives could be combined with melodrama as a means of self-expression. Banta points out that muckraking journalists engaged in a deliberately divided discourse in which "narratives of injustice spoke in the name of scientific objectivity while calling up the wild facts of the heart" (24). If Taylorism was rapidly seen as a "syste[m] of control," as Banta asserts (26), melodrama might have been the mode best suited to detail the excesses of that control and to escape them. Yet, like the muckrakers' accounts, melodramatic screenplays also contained the "professional" elements necessary to give them plausibility. In other words, screenwriters were not necessarily being steered away from melodramatic devices so much as being asked to motivate them acceptably (in Taylorist terms, make them "efficient") and to present them in the modern, streamlined form associated with professional scriptwriting. Thus Palmer could offer its students a fully modern tool for self-expression, which, while partaking of the regimented character of the office and factory through standardization, nonetheless allowed them to express their disenchantment with such environments through melodrama.[9]

Advertising and Self-Commodification

Palmer was very careful not to promise material success to subscribers, but rather to sell the less tangible advantages of the developed personality. By fostering narrative ability, the Palmer Plan offers a new self, as Roy Manker, in 1925 the company's vice-president, points out:

> The highest achievement of mankind is not a million dollars, but a
> personality. The ability to see and tell a story contributes immensely
> towards the formation of personality.... It is a realization of this that is
> awakening more people to the value of knowing how to talk and write

well, than ever gave the subject a thought in the past. . . . Every one wants to be set off a little in the crowd, and he recognizes that the readiest means to that end is the practice of the short story teller's art. (Quoted in Wayne, 258–59)

Not only is this emphasis upon personality good business sense (would students ever complain that their personalities remained flaccid and unappealing even after rigorous study?) but it also immediately broaches the project of self-commodification at the root of the Palmer ethos.

This emphasis on the formation of personality through fiction writing implies that the Palmer Plan can be viewed within the context of the larger culture of personality; this culture operated, argue Warren Susman and T. J. Jackson Lears, from the late nineteenth century onward. According to Susman's analysis of the terms associated with personality in advice manuals of the 1880s through the 1920s, the culture of personality emphasized the "masterful, creative, dominant, forceful" (277). Those who had personality could stand out from the crowd—à la Ruth Fielding, perhaps—different yet at the same time appealing to the general taste. Lears notes that "as the self became commodified, it was also expressed more often through the consumption of commodities" ("Salvation," 37). In the case of the Palmer Photoplay Corporation, the course appeared at first blush to be the commodity, but its ultimate aim was to help the student perfect the self as a commodity, either by fitting herself or himself to an entirely new career or by acquiring a training that would make the self intrinsically more desirable. In either instance, the student was being prepared to create a self that could circulate among others and that was, on some level, the product of internalizing other people's views of what that self could be. As with the films and exhibition practices described in juvenile series fiction, the measure of virtue for both scripts and personalities in Palmerdom is the extent to which they meet the approval of the (nonscriptwriting) masses. Each sign of approval—a theater admission paid, a party invitation extended—is a vote of confidence.

Democracy thus becomes rhetorically crucial, although this rhetoric is not, in fact, uncomplicatedly democratic. In an attempt to steer the design of the course between the shoals of conflicting representations of authorial success as the product of labor (a democratic conceit) or of talent (a potentially more elitist conceit), Palmer had to represent success much as the authors of juvenile series fiction had represented it— that is, as the result of latent ability rigorously developed. This approach restored what might otherwise be lacking in the new culture of personality, equality of opportunity within an industry otherwise appearing exclusive through its privileging of looks, talent, or hard-to-obtain experience. Here, discipline and the promotion of conformity to standards come to Palmer's aid. As Walter Benn Michaels has argued in the context of Progressive Era literature, "Individuality now appears as an effect of standardization" (73). Michaels cites John William Ward's notion of "the shift from a nineteenth-century conception of individualism that, at its most extreme, declared the individual an individual only 'to the extent that he was separate from society' to a twentieth-century conception that, as *its* extreme, identifies individualism as the 'submersion of the individual in the group'" (quoted in Michaels, 72). The quality of being an individual emerges through the adoption of methods designed, ironically, to apply to anyone who wishes success. Palmer's standards, in other words, function to highlight and reveal the individuality already latent. Adherence to standards is the way to create a desirable product from the self, which, at once agent and patient in this transaction, is already "fascinating" and simply needs to be marketed as such.

The notion of the already potentially "fascinating" self is where Palmer's advertisements are most manipulative. The desirable characteristics innate in the ideal applicant are described in such mystical terms that they seem not to correspond to any particular personality type or state of preparation, but instead constitute the yin and yang of scenario writing:

> *Dramatic Insight* is that power of mind, that inner vision which enables whoever possesses it to see in all forms of life, from the humblest to the most exalted, dramatic situations, or the possibilities for them. It is the power that draws to itself, as the magnet draws the needle, all that is thrilling in any relationship of characters, in any situation which pulls at the emotions.
>
> *Creative Imagination* is the active counterpart to this passive power. It takes all of the things that the dramatic insight draws to itself as raw material, and by a process of development and arrangements shapes these facts of life into a finished structure, a photoplay of power, penetration and distinction. (Palmer, *Essentials*, 10)

In this highly charged copy, readers are addressed as if already what they hope to become through study—alluring and powerful. And while these talents may be inherent, they are by no means the province of an exclusive few; on the contrary, they "are commoner than is generally supposed" (Palmer, *Essentials,* 10). Through its promise to hone what already exists within the student, Palmer's advertising literature sets up a no-lose proposition.

Early Palmer advertisements employed rhetorical devices nearly identical to those of Lewis's fictional example, trading not only in the tropes of "fascination" but also more crassly in power and money. An advertisement in *Photoplay* elects the "power over others" motif, with a headline that inquires, "Who told Chas. Ray [an actor providing a testimonial for the corporation] to act this way?" (February 1919, 96). In other words, it is not in acting that the mastery possible in Hollywood films lies; the real power exists out of the public eye, but the astute Palmer graduate may well become part of that inner circle. In suggesting the separation of planning and conception from the world of visible labor (acting), Palmer confirms its graduates in an essentially Taylorist view of filmmaking. On a more mercenary note, other ads emphasized the monetary benefits of the Palmer Plan ("not a mere book nor a 'school,' nor a tedious correspondence course"), suggesting that one might make between $100 and $1000 from a properly presented photoplay idea (*Photoplay* April 1919, 98, and May 1919, 98).

Later advertisements, such as that published in *Writer's Monthly* in March 1925 (275), suggest that, by the mid-1920s, the corporation's focus was shifting from scenario writing to more general literary pursuits, a widening of scope that no doubt reflected the constriction of the freelance writing market. The possibilities represented by such a market were, however, kept alive not only by Palmer but also, into the 1920s, by fan magazines, which published articles suggesting that the deserving might indeed have their place in the film industry. Some of these pieces were nonfiction, such as the 1918 *Photoplay* series by Anita Loos and John Emerson offering practical advice on photoplaywriting (subsequently published as *How to Write Photoplays* [1920]); others were fictional treatments, such as Frederick Arnold Kummer's "The Rejected One" (1918), the story of a young man who repeatedly tries to sell his work to a studio and finally gets past the Cerberus at the scenario editor's desk by putting his work inside a cover with a well-known author's name on it.

Such stories and articles necessarily focus substantially on the personality traits and background desirable for the successful screenwriter. In advertisements, Palmer students were encouraged to think of themselves as simultaneously the raw material for potential success and the agents acting upon this raw material. The rhetoric of self-commodification urged students to enroll to learn "how to put your movie ideas into actual *cashable* form" (*Photoplay* May 1919, 98). An ad depicting a despondent author slumped over his typewriter, rejection letter before him, catches the eye with the headline, "And he had put his whole soul into that story!" and the cure for editorial neglect is to "learn how to stand off from your work and view it coldly and critically" (*Writer's Monthly* July 1925, 83). A certain alienation of the self, in other words, seems embedded in this process of circulating profitably among other people, as is also true for Ruth Fielding in the juvenile series fiction. For both Ruth and the Palmer students, self-commodification is the way out of wage slavery. If the amateur simply masters the art of writing desirable screenplays that mine the self for valuable material, then that material, rather than one's own labor, can be marketed.

In a Palmer pamphlet offering the potential customer *Little Stories of Success,* wage slaves were precisely those solicited:

> Why pour your energies into a task that is distasteful, when by using a part of your spare time you may prepare for a position worthy of your ability? Why remain at an inferior station and watch the rest of the world enjoy advantages and possessions that you too may share? Why strain your energies to make adequate a mediocre income when you may have the equal of your present whole year's salary in one successful photoplay? (16)

Yet even Palmer was obliged to hedge its bets through the recognition that the structure of the freelance market in Hollywood was changing. The Palmer literature acknowledges this alteration by suggesting employment in "high salaried positions as staff scenarists" as an alternative (or addition) to the achievement of "large incomes as free-lance writers" (*Little Stories,* 16), although the freelance option is clearly more attractive to student and Plan alike. The Palmer course serves as the sympathetic and knowledgeable guide that would enable the student "to profit from the undeveloped resources of [his] own imagination" (Palmer, *Manual,* ix). Not only is self-commodification through the mastery of screenwriting technique profitable, but it also appears to reconcile some of the instabilities latent in the opposition between creative individual and standardizing system.

If standards were the way to become more individual, then, they were also the way to ensure democratic access to a desired field. Samuel Haber notes that Frederick Winslow Taylor "fashioned his methods after the exact sciences—experiment, measurement, generalization—in the hope of discovering laws of management which, like laws of nature, would be impartial and above class prejudice" (x). While no one pretended that screenwriting was an exact science, Palmer literature presented the film industry's standards for good storytelling as part of a carefully engineered system (which they were). The democratic appeal was the same that endeared Taylorism to Progressives generally, namely that standards were knowable, predictable, and above claims of privilege.[10] Mastery of the right set of standards (significantly, *not* those of highbrow endeavors) was open to all. Palmer personnel made this last point clear. In "Functions of the Continuity Writer," Jeanie MacPherson (valued assistant to Cecil B. DeMille) scotched the notion that privilege and success in other areas necessarily provided entrée to even the lowest rungs of this new field:

> "But it's just 'hack' work, isn't it?" she asked, and there was a plaintive, injured note in the voice of the earnest would-be screen writer who had applied in vain to many studios, requesting a "position as a staff continuity writer, while learning to write photodramas"!!! Strange to say, the questioner was no illiterate blunderer, but a college graduate and a brilliant writer in the realms of philosophy and poetry. (25)

If one apprenticed oneself to Palmer, however, one received all the training one needed—success followed the rigorous applications of Palmer standards.

And for those who feared that their raw material might be a little lacking, and thus not amenable to improvement even with the most careful instruction, Palmer literature pointed out that it was the least intellectual, most innate aspects of personality that contained the germ of photoplay success. In other words, it suggested that the terms "highbrow" and "artistic" be uncoupled. In "Study Your Audience," Harvey O'Higgins invoked Freudian terminology to reassure his readers that they already possessed the real thing:

> The subconscious mind is the mind that dreams while you sleep. It is the mind in which your animal instincts move. It is the primitive mind out of which conscious intelligence has developed. It is the mind of the instinctive emotions. And it is the mind to which every successful story and play and movie has to go to get the emotional pull that "puts it

over." It is the artistic mind, as distinguished from the scientific mind. It is the lowbrow mind, not the highbrow. It is what we call "the heart," as contrasted with "the head." (9)

In this representation, as in much else, Palmer rhetoric allowed the reader to have it both ways. On the one hand, dreams were just the raw stuff of film and thus available to all, and on the other, understanding this truth involved at least a nodding acquaintance with new psychological concepts, the privileged "scientific" knowledge in this article. Screenplays themselves were not coldly analytical, although the author might have to be to excavate the passionate material lying inside his or her psyche, in the perfect marriage between the popular touch and modern technocracy.

The device that most powerfully suggested that the screenwriting field was open to all comers was the photoplay contest to "discover" new talent, common throughout the 1910s and early 1920s (and memorialized in *Ruth Fielding and Her Great Scenario*). As early as 1916, a trade paper commentator labeled such contests mere advertising stunts and not the entrée into a career they were sometimes represented as (Wright, 43), but they nonetheless remained the preeminent way of soliciting public participation in Hollywood. Contests, like standards, fostered the double rhetoric of film as both a democratic enterprise and one that offered the potential for individual elevation above the crowd. In April 1922, Goldwyn Pictures (on the verge of abandoning the Eminent Authors for more obscure talent) and the Chicago *Daily News* jointly solicited manuscripts in a contest with a top prize of ten thousand dollars; it attracted twenty-seven thousand submissions ("Florida Woman," 721). Winifred Kimball's *Broken Chains* won first place, and Kimball was claimed as a Palmer graduate in a July 1922 *Photodramatist* advertisement that observed, "more than $20,000 of the $40,000 offered in prizes during the last year or so has been won by Palmer students" (4).[11]

Even in the absence of an open freelance market, then, the Palmer Plan might represent a good investment if one were fortunate enough to win a contest. The celebrity associated with such a victory continued the confusion between person and product that was so intrinsic to the project of self-commodification. For example, in what turned out to be a persistent slippage between self and commodity, Kimball's personal narrative of success was conflated with the narrative sold for production as a film:

These three years [of Kimball's apprenticeship with Palmer] were no
easy period of casual study. To have the dark tragedy of suicide twice
enter one's home, to be forced into reduced circumstances with others
dependent on one for support, to have the high faith and courage to
spend money and time from a slender store on the study of dramatic art,
to persist gamely through three years of discouragement while things
grew steadily worse and the mortgage on the old home became past due
and then when things were the blackest, to win enough to clear everything, in competition with 27,000 other contestants, such is the history
in brief of Miss Kimball's rise to fame. It is a narrative of fact with a
dramatic climax only equalled by her photoplay story "Broken Chains."
(S. Johnson, 21)

It is hard to say whether the study of technique or the endurance of
hardship had more to do with the success of Kimball's work, according
to this account, because the reader learns of her sufferings almost as if
they were the subject of her studies. Certainly her success is the happy
result of combining suffering and study, suggesting that turning one's
pain into affect and thus into a marketable item was important to the
project of self-commodification; presumably the suffering that Kimball
endured turned her into a better, stronger human being and thus a superior artist, even if *Broken Chains* did not directly dramatize the suicides,
the penury, and the threat of eviction from "the old home."

In line with this commentary on Kimball, Palmer writings reiterate
that a key qualification for screenwriting is experience. Defined in one
way, "experience" could mean practice in genres considered akin to screenwriting, such as journalism. Indeed, Palmer claimed that customers could
leverage the knowledge gained in some other occupation into success in
screenwriting. Don Ryan reported that the corporation solicited him as
a subscriber precisely because he was a journalist;[12] the form letter, signed
by Manker, read in part:

As a newspaper man on the desk and reportorial staff of dailies in
Philadelphia, Boston, Chicago and other cities, I always cherished the
idea of some day doing "bigger and finer things," as Merton of the
Movies would say. Short stories and magazine work interested me—
then scenario writing claimed me. My newspaper experience offered me
so much story material to build upon that nothing but the mastery of
scenario technique remained to be acquired. The combination spelled
success. (Quoted in Ryan, 8)

remarkable marketing strengths, which were hardly personal, were fundamental in putting the products of the Palmer Plan before the American public.

Precisely because the Plan was the artifact of a corporate undertaking, it could offer students what appeared to be individual attention. The sales department, headed by Kate Corbaley, herself the quondam winner of a photoplay contest (Denton, 83) and later story editor at MGM,[13] affords an example of the numerous ways Palmer was able to present itself as a corporate enterprise working in the interests of specific individuals. Not only would the department serve as agent for qualified students, it was available in this role also to professional writers willing to pay a twenty-five dollar fee annually, thus implying that unfledged students were already part of the august company of authors. The department advertised that it kept its finger on the pulse of the business by performing the scientific study of market conditions that most students could not do for themselves. Evidently, scripts were often submitted with specific actors in mind, so that the author needed to know where a particular star might be working and what kind of stories interested him or her. Corbaley observes:

> It is quite impossible for authors living at a distance to keep a record of these changes [in actors' employment or desired subject matter], for by the time they are reported in the trade publications a week or a month may have elapsed and still another change may have been made. . . . It is the business of the Palmer Sales Department to keep in daily contact with these fluctuating activities and maintain an accurate record of all changes that occur. (5–6)

The personal touch available within the confines of a large corporation also arose from the sales techniques used by Palmer; stories "[were] not mailed promiscuously to one studio, and then, if rejected, to another," but rather hand-carried by a Palmer representative and discussed with a scenario editor, producer, director, or star (9).

In its guise as agent, the sales department enforced uniformity of appearance for every manuscript sent out for consideration. The process, as Corbaley describes it, was designed partly to give each story polish and partly to discourage plagiarism by unscrupulous studios:

> Stories passed upon favorably by the Sales Department are sent to the typing room to be copied on our special manuscript paper. This paper is steel gray and each page is outlined in blue lines; the typing is done in

blue ink of the same color as the ruling and each page carries at the
bottom in small blue letters the name, Palmer Photoplay Corporation.
These pages are placed in a rough gray folder, bound into book form, the
cover bearing the print of the seal of the Corporation in blue and red. . . .
Thus the Palmer Photoplay Corporation presents the scenario-synopsis
in conjunction with the author.[14] (Corbaley, 7–8)

This treatment of the salable submission in effect mass-produced manuscripts. Agented, retyped, protected from plagiarism through its association with the corporation (which registered each scenario submitted
to studios), the student's idea became physical artifact, commodity, and
part of a national brand so that it could achieve success on its merits,
much as the hero of Kummer's story (quoted above) discovered that
tucking his scenario inside a folder bearing a well-known name would get
the manuscript the reading denied earlier. Moreover, in keeping with
other corporations developing national brands, Palmer united a dispersed
workforce into a single production operation, extolling its instruction
department as "a necessary adjunct or 'feeder' for the sales department,
to which it looks for its chief source of revenue in the future" (Palmer,
Essentials, 31).

Significantly, the sales function was reserved to the corporation, and
the student was discouraged from taking it up: "The author must be well
equipped with knowledge of photoplay plot technique, but it is equally
necessary that the selling agent base his efforts on a market technique
that may be learned only after intensive experience and a definite study
of the many angles and details of selling" (Corbaley, 9). At this point,
the successful student was producing wealth both for her- or himself
and for Palmer, which took a ten percent commission on any sales
through its sales department. The student was not permitted to forget
that motion pictures are a collaborative art and that her or his achievement therefore depended to some extent on the work of others, as *The
Essentials of Photoplay Writing* pointed out: "When you submit a photoplay synopsis you have the assistance of staff scenario writers, directors,
actors, scenery and expert photography to amplify all the little details of
production" (Palmer, 11). This invocation of the group effort helped to
establish screenwriting as simultaneously democratic/cooperative (with
power coming from uniting the mass in a single glorious enterprise)
and industrial/authoritarian (with scenario writing akin to the work of
a business executive assisted by a large staff of underlings). On one level,

then, the scenario worked within a discourse of mastery in which, according to Palmer, the important brain work of the Taylorized studio was reserved to the student, while the more tedious details were left to the "assistants" employed by Palmer on that student's behalf.

Collaboration occurred in other spheres of photoplay labor. If the student found it disheartening to work alone until success arrived, he could read the advertisements suggesting collaborative effort in his monthly *Photodramatist*. Beyond this, he could join one of the many groups of Palmer students springing up around the country. At first, *Photodramatist* greeted these groups as an excellent idea, when a student named David Bader wrote in to thank Frederick Palmer for his offer of help in setting up such a club. This club's program was to read a chapter of the Palmer Plan, discuss it, and then analyze recent films (13). Apparently club meetings could become confusing or acrimonious, however, for *Photodramatist* counseled against them three months later, arguing in July 1921 that working alone resulted in greater originality. Given that the construction of a new self is no easy task, what was even worse than the spread of technical errors through the absorption of misconceptions disseminated by fellow students was the communication of despair:

> In one Palmer Club that disbanded the disintegration process started when the vibrations of one hapless little discourager began filtering around. Almost every member of the club submitted at least one manuscript to the Palmer revision department. Every one was returned. One or two gloomy mortals did the rest, and almost every member of the "club" began to believe that he or she lacked any talent, that the future was hopeless. (LeBerthon, 13)

While students might do well to study together, it was clearly not in Palmer's best interests to promote comparing notes about relative abilities or rates of success.

Indeed, on the same principle Palmer had to urge already enrolled students not to fill out the questionnaire used to divine the talent and readiness of the not-yet-enrolled. The rhetoric of the brochures, like that of the advertisements, walked a fine line between inclusiveness (anyone can learn to write photoplays if properly taught) and exclusiveness (only those with some native ability, even if presently obscure, can follow the Plan to success). This rhetoric was shaped, in part, by the exigencies of the business, which could not safely appear to encourage the hopeless simply to extort more money from them. Just as important, however, was that this rhetoric resulted from the same uneasy relation between individual and crowd that produced some fraction of the audience for correspondence schools in the first place—the desire to distinguish oneself while remaining in touch with those one hoped to impress or dominate. The crowd of which one was a member had, at least, to be an elite crowd, and Palmer students sometimes had to be reminded that they wrote for the entertainment of the masses, not for an audience of a select few (see, e.g., MacPherson, *Necessity,* 8–9).

It is interesting to contemplate such (unreliable) membership figures as are presently available in light of this tension between elite author and more plebeian audience.[15] Palmer invited current members to sign up family and friends (for instance, in an advertisement on the back cover of *Photoplaywright* in January 1921), thus encouraging enrollment; this invitation certainly suggests inclusiveness—and possibly a willingness to commodify others. At the same time, however, the corporation was at pains to suggest that many more potential students were weeded out by the admission questionnaire than were permitted to enroll. *Photodramatist* reported in 1922 that

> in this year, between January and July 1, 4301 of the people who applied to the Palmer Photoplay Corporation for the psychological test which serves as their entrance examination, were rejected. Out of every 100 applicants, only 12 are invited to enroll and only three out of 12—in other words, three out of the 100—pass with distinction. The other nine are told they can take the course on their own responsibility and are warned that the way will be hard. (Stevens, 18)

Similarly, *Moving Picture World* listed among Frederick Palmer's successes the fact that he had "turned down close to 50,000 applicants" in less than five years (Tidden, 553).

Nonetheless, in 1921 Palmer framed his mission in terms of the mass production of manuscripts, needed to fill the demand for acceptable stories:

> Gradually the realization came to me that there lay before me a mission in life that might be more useful and more far-reaching than the writing of photoplays; for if I could reach and teach others to do what I was doing I could indirectly be the cause of producing a far greater supply of manuscripts than any single writer or limited group of writers could ever hope to do. (*Essentials,* 5)

Palmer himself is frequently described as having written fifty-two scenarios in nine months (*Essentials,* 4), entitling him to speak of screenwriting as a form of industrial production (although his record was established in 1910–11 when scenarios were substantially shorter). As he delicately notes, he was the indirect cause of more manuscripts; however, the quoted passage glosses over the most important industrial production that the school engaged in, the manufacture of the students who wrote the photoplays.

Palmerplays and the Individual

The enrollment figures, which seem too good to be true, were offered in autumn 1922, when Palmer was expanding into film production—the crowning moment of collaboration and, incidentally, further vertical integration on the part of the corporation. Now the school would instruct its students *and* critique, agent, and produce their works. Because Palmer had many mansions, so to speak, it could be inclusive and exclusive at the same moment; many more would enter the school than would ever complete a scenario, let alone sell one, and fewer still would have one produced by the school itself. The student could nonetheless feel pride as he or she passed through each wicket on the way to greater recognition, and even the unsuccessful could feel that they had gained something through their more fortunate fellow students. At the time of the release of its first Palmerplay, the corporation's students were urged to write to their local theaters to ask that the film be booked there, thereby contributing to the film's commercial success and throwing off the yoke of the Eastern writing establishment by sponsoring their own products in preference to adaptations.[16] In the same circular, Manker directly invokes the notion of the plebiscite at the box office in suggesting, "If we are right in assuming that you rebel at being taxed without representation, then cast your ballot for democratizing the screen by writing a letter to the manager of your favorite theatre requesting him to exhibit this picture" (undated circular letter to subscribers). This particular rhetoric of democracy urged students to experience success vicariously, if need be. Manker asserted of *Judgment of the Storm* that "it is your picture as much as the author's or ours, because it was written by one of your fellow students and produced by our institution," suggesting that despite the emphasis on individual professional achievement, a main aim was to foster a sense of kinship among participants by allowing them controlled arenas for concerted action.

Yet the increasingly complete incorporation of student work in the corporate enterprise suggests that recognition in this highly collaborative medium may have had its price, because while greater and greater mastery of self-expression should have led to the further development of the self or "personality," it may also have had the effect of submerging it, at least to some degree. The student succeeded, presumably, because he or she had heeded the guidelines, mastered the jargon, and sent out a manuscript flying the Palmer colors. But even if the story was produced, some doubt as to whose narrative it finally was may have lingered. Take, for instance, *Variety*'s generally glowing review, in its 31 January 1924 issue, of the first effort in the Palmerplay production line:

> "Judgment of the Storm" if exploited properly should have no trouble cashing in. It is a Palmer Production produced by the Palmer Photoplay Corp., made up of former newspaper men, evidently with plenty of money and who threaten to be heard from if they keep up the same pace.
>
> The story is a first effort, written by Ethel Styles Middleton, described in the billing as a simple and domestic housewife of Pittsburgh. From the workmanlike construction of the story and the many old but always effective and clever tricks of the trade utilized, it seems likely that the "housewife" business is merely bait, or possibly that some experienced, crafty rewrite man fixed the plot up to suit himself.

Ironically, the very methods designed to bring success to the novice scenarist also worked to undermine his or her individuality.

Another irony is that if the three Palmer productions reviewed in *Variety* have a group flavor, it is their tendency toward melodrama, often a source of anxiety for industry commentators (such as the educators discussed in chapter 5), and an element that the Palmer textbooks urged students to use only sparingly. Nevertheless, the Palmer curriculum forgave the employment of melodrama when properly justified—since the control and expert manipulation of affect were the key to the mastery of medium and audience alike. As Sheldon Krag Johnson comments approvingly of Kimball's *Broken Chains:*

> Is melodrama the portrayal of physical jeopardy? If so, if thrills, strong situations and repeated incidents of physical peril are melodramatic, then this play is indeed a dark swift stream of melodrama. But if

melodrama is the depicting of strong situations and violent acts which
have no proper justification in the motivation of the characters then
"Broken Chains" is by no means a melodrama. (21–22)

Clearly, if we use these standards, *Judgment of the Storm* features melodrama both good and bad, with *Way Down East*–like moments that rely
on a terrible storm for some of its most stirring effects (shades of Ruth
Fielding). Wrenching decisions made in the face of nature's fury combine both physical and mental "punch," which suggests an interesting
distinction in the way that affect is ideally produced: mental punch is to
be preferred to merely physical punch, which corresponds more closely
to insufficiently motivated melodrama.

As Palmer notes, "there is a physical punch in hand to hand combat;
in a hunter facing a ferocious lion; in a girl marooned on a tenth floor
fire escape of a burning building," but this technique pales beside mental punch, which "involves reaction that is mental or moral" (*Self-
Criticism*, 8–9). *Variety* lauds the combination of physical and mental
punch in describing the film's climax:

> There follows another big scene, possibly unreal but extraordinarily
> gripping. John's mother, dying from exposure and realizing he cannot
> save all, commands him to take the twins and return them to the home
> so recently saddened because of her. He obeys but returns in time to
> rescue her too and the story ends in a halo of sunshine.

Presumably the mental punch (the necessity of deciding between mother
and twins) redeems the physical punch of the storm itself, although
clearly the devices are thoroughly intertwined. The distinction between
mental and physical punch nonetheless suggests that scenarists and reviewers thought that affect had to be created within the parameters of
the realist mode.

Variety's reviews of the second two Palmerplays, however, demonstrate how recourse to formula may undermine the individuality of a film
(and suggest that Palmerplays did not always practice what the Palmer
course preached). Palmer's second film, *The White Sin*, is described more
temperately as "swift moving, dramatic and convincing except in one or
two spots. The author found it necessary to work in a fire and rescue to
plant his hero and heroine for the final embrace" (14 May 1924). But by
the third Palmer release, *His Forgotten Wife*, the reviewer becomes positively impatient, declaring the film "another of those amnesia and long

arm of coincidence pictures, filled with the ancient hokum of the film business, to wit, two women after one man, automobile chases, the daring of the good woman to help the man she loves, the old family lawyer, and numerous clinch scenes" (25 June 1924). As in juvenile series fiction, one notes a delicate balance between the necessary understanding of the conventions or rules of film narration (which include not only continuity but also certain archetypal situations) and the imperative to produce entertainment that can be considered novel. If one fails to strike this balance, the attempt to produce and market affect becomes merely annoying (if too clichéd) or incomprehensible (if too bizarre).

The reliance on formula for success seemed to undermine not only the production of personality (raising the question of *whose* the successful picture was) but also the avowed purpose of the corporation's move into film manufacture: to showcase fresh authorial talent and to reward the independent author through royalties.[17] Indeed, the best way of describing the Palmer foray into the making of actual movies is to liken it to a kind of Eminent Authors program for unknowns. Poised between warring approaches, the Palmer film production project exposes the awkwardly opposed aspects of the rhetoric surrounding the film industry. On the one hand, film is a democratic field and anyone can enter it; on the other, it requires both training and talent. It is a failing of the amateur to fall back on the creaky devices that were old when stage melodrama was young; nonetheless, the experienced staff approving pictures for production appears to select just those stories with the hoary aspects that so offend *Variety*'s reviewer. Authorship is about lone creativity, some may say, although the film production process requires the collaboration of many, and the reviews of these pictures chiefly imply that the good, hard-working casts are putting the weak stories over.

Finally and perhaps crucially, the Palmerplays intervene directly in the debate raging over film's destiny as either adapter of material from other sources (novels, the stage, magazines) or producer of original narratives. The value of the pre-sold property was a major topic of debate and analysis in the early 1920s, and it was often attributed either to the structure of film distribution or, in some instances, to the low caliber of audience taste. Because film was so expensive to produce, it had to exploit reliable and already extant markets. For example, Tamar Lane noted in 1923 that when [a producer] . . . offers \$5,000 or \$125,000 for the rights to some noted play or book it is not because the story has any particular merit. It is purely because he feels that the particular story will draw big gate receipts. Many are the works the movie impresario buys at high rates of which he has never even read three lines. But he has heard [them] talked of in public and that is sufficient. (42)

This debate was not merely a feature of the trade papers; Palmer students were obliged to contemplate this aspect of film production through their reading of *Photodramatist*. Like Lane (who wrote for *Motion Picture Magazine* and *The Screen*), Rob Wagner deplored the expense of big adaptations and the recourse to stories culled from other media as a hedge against failure. Wagner reports an imaginary conversation between the production managers in California and the distribution managers in New York, with the latter saying, "The story is good and we feel that it will be a splendid vehicle for you, but can't you induce Wagner to shoot it into a magazine as our salesmen can get a much bigger price from exhibitors if the story has first appeared in a national magazine?" (3). Wagner bemoans film's reliance on unappealing or unseemly works from other media. He presents arguments on both sides of the divide, with the evidence more or less weighted toward the pre-sold property by his understanding of the structure of film finance and distribution. Nonetheless, he concludes that originals will become the wave of the future because they must; film's high destiny as an art form requires this (3). He thus returns to the imputation on which Palmer's fortunes were based, that fresh blood is essential to the continued creation of the Hollywood product.

The debate was not merely about suckering students into believing that they might enter an industry that had become, by the early 1920s, as hierarchical and closed to the unconnected, untrained, and naive as was any other. Rather, the discussion manifested discomfort with certain structural aspects of the film industry at the highest levels, a discomfort that Will Hays expended much effort trying to assuage, often in rhetoric as strained as any appearing in Palmer publications. While the pre-sold property was a rational attempt to defray the costs of production and distribution by capitalizing on the familiar, it involved the film industry directly in invidious comparisons with the publishing industry, which had constitutional protections that film lacked. As Richard

Maltby observes, Hays's avowed desire to "prevent the prevalent type of book and play from becoming the prevalent type of picture" was not the outward and visible sign of an inward and spiritual failure of taste, but rather a recognition that, without constitutional protections, the film industry could only adapt works from other media in such a way as to minimize objectionable features; forgoing adaptation altogether was not an option because of the perennial shortage of material and the box-office leverage of the pre-sold property ("'Book,'" 100–3). After 1924, the film industry employed the "Formula" (typically honored more in the breach than in the observance) designed to control adaptations from other media. The presence of the "Formula" nonetheless clarified film's debt to publishing and the stage; it was not a declaration of independence.

Its emphasis on the rhetoric of independence from the publishing establishment notwithstanding, Palmer, through its strategy of "novelizing" *Judgment of the Storm,* participated in a form of vassalage to the publishing industry. Doubleday, Page, and Company published the novel version at the time of the film's release in January 1924. This edition of the photoplay contained not only the novel proper but also the "cutting script" (hailed as a novelty) and stills from the film, permitting the publisher to call it the "professional" edition. Needless to say, Palmer urged its purchase on the grounds that it would make an admirable textbook for the aspiring scenarist; again, the way to take bold new steps in changing one's career, and by extension one's self, was through imitation.

But the cutting script was also described in such a way as to urge taking up screenwriting as a form of artistic mastery (see "Send No Money"):[18]

> This cutting script unquestionably represents the most important and fascinating phase of film production. It regulates the editing of the film. It exercises judgment over the story, direction, photography, research and every detail involved in the production of the picture. It packs dramatic action, recorded on miles of film, into a few thousand feet.

In short, the script itself is presented as a hero of the personality age— fascinating, efficient, authoritative, dynamic, everything the Palmer adept would hope to be. Significantly, it can be possessed. It is a physical artifact from a culture whose creations are usually ephemeral, and this is another point the advertising takes pains to stress: "The story is a faithful, permanent elaboration of the photoplay which, unfortunately, must flit across the silver screens of the country and be lost" ("Send No Money").

Moreover, this hybrid work of fiction illuminates the vexing problem of authorship when the power and aims of a corporation are advising and directing the would-be author. Middleton is not the "author" of this work in its permanent physical form. Instead, the author of *Judgment of the Storm*, the novel, is Roy Mason; thus novelization may undermine the newfound screen author in favor of the more experienced literary hack. In addition, the strategy of the novelization is in line with the industry's desire to limit risk through the exploitation of pre-sold properties. The novelization turns Middleton's idea into a pre-sold property in which her name is not the most significant. Indeed, Palmer students are urged to read the book before they attend the film, thus making the film a pre-sold property after the fact.

Even pre-sold properties and the enthusiasm of the student body were not enough to keep the production project afloat after the summer of 1924, and it is possible that Palmer entered film production largely to shore up its market as a correspondence school, as *Variety* and Sargent charged.[19] The Film Booking Office distributed the pictures, which seem not to have made a big impression on the industry after the first release. While the early- and mid-1920s saw substantial economic consolidation in the film industry, particularly in exhibition and distribution, the market still had room for small independent producers. These conditions were gone by the end of the decade, but in the early 1920s Palmer might reasonably have thought that film production could bring benefits beyond increased enrollment or the gratification of those already enrolled. In any event, Palmer could argue that it was already providing sales opportunities to its students by allowing them to place short story versions of their photoplay ideas in the house organ, known as *The Story World and Photodramatist* by the time the film production scheme was launched.

Because of Palmer's protean nature, it is difficult at times to identify exactly what it was selling, a feature that its thoughtful choice of rhetoric only intensified. The firm may be said to have hawked variously a mystique, an adjunct education, a set of experiences, a way of commodifying past experiences and turning them into ready cash, mastery over the self and others, membership in an elect community, filmgoing, specific films of its own manufacture, and accessories. The importance of this last item is demonstrated by a startling description of the efficacy of another commodity, the *Photoplay Plot Encyclopedia,* which student and produced author Martha Lord praised in terms as appropriate to a drain cleaner:

> In plot construction the Palmer Photoplay Encyclopedia is invaluable,— really, I have been able to work out the most unique situations through its aid. I keep it constantly by me and whenever I get my characters into such a predicament that an easy death seems the only solution of the problem, I delve into its pages, and presto! the plot uncoils. (Quoted in "Inner Sight," 7)

Given Palmer's definition of a photoplay story ("getting the characters into a predicament and then getting them out of it" [Palmer, *Handbook,* 1918, 58]), the *Encyclopedia* was clearly an essential tool. In fostering the appeal of physical artifacts associated with film, the corporation, like the film industry at large, promoted a deliberate confusion between being and having, doing and appearing. If one needed such tools to make a photoplay, who was performing the labor—the author or the infallible plot decongester?

The *Encyclopedia* functioned as a representative of the world of professional filmmaking, a guide to what was going on behind the scenes, and a purveyor of technical vocabulary. If the correspondence schools overstated their ability to aid in any material way, some students may have been satisfied with the possible rise in status that merely claiming connection with the enterprise would secure them. The desire to appear knowledgeable about technical aspects of the film world must have been widely dispersed in the 1920s, to judge from the extraordinary number of publications from this period that include glossaries of film terms for the lay person, and mastering this new technological language might, by itself, have been enough for some subscribers.

If technical vocabulary alone did not sate one's appetite for learning, the remarkable range of reference employed in the various Palmer lectures suggests that a desire to complete an inadequate education was another strong motive. For example, Clarence G. Badger's *The Point of Attack, or How to Start the Photoplay* invokes Gustave Flaubert and Alexandre Dumas *père* as exemplars or purveyors of advice. Denison Clift's *Dramatic Suspense in the Photoplay* employs Gustav Freytag and Henrik

Ibsen. When we consider that these lectures are larded with references to photoplays such as *The Miracle Man* and Corbaley's *Gates of Brass,* we understand that part of what is at stake in the Palmer Plan is the ratification of filmgoing and of the filmgoer's taste. Through his or her love of film, moreover, the interested student can painlessly acquire a nodding acquaintance with the literary giants of the past, whose works are effectively being attached to film culture in these manuals. This strategy prefigures a desire to foster a love of the classics through film attendance (see chapter 5). Finally, when the film industry lives up to its high destiny, it will produce works that can be mentioned in the same breath as those penned by the most august and daring nineteenth-century masters.

Such references to the peaks of literary art were the unexpected verso of a program with aims frequently vigorously pecuniary; as Frederick Palmer pointed out, the job was "to write and SELL stories" (*Handbook* 1918, 18).[20] By offering a smattering of literary aperçus and psychological insights, the Plan appeared to have something to benefit the individual and the industry, even if not every student could hope to become a professional author. Given the frequent recourse to notions of audience inadequacy as an explanation for film's failure to rise to the artistic heights predicted (see chapter 4), some program of "audience education" would have appeared well advised. Writing for Palmer students, Douglas Z. Doty noted:

> It is significant that so many people are engaged in attempting to write for the screen; and already there has been a curious reaction. Few of these writers are commercially successful, but it is distinctly noticeable that the average motion picture audience is growing much more critical of what it sees, and this is bound to have a salutary effect on the productions of the future. (20)

More precisely, *Photodramatist* advised Palmer students that, "although the student should fail to sell a single story, the broad mental training received in studying the course should prove of advantage in meeting the world at any angle" (Moracin, 15).

But to turn one's marketability to good account, the Palmer literature argued, one had to join a crowd of like-minded people. In fact, only by joining the ranks of Palmer students—by unionizing, in effect—could one be assured of a fair shake in the battle to enter the film industry. Not only would one learn what the Palmer Plan had to teach, but one would benefit from the features of the Plan especially designed to gain entrée for one's work, particularly if one knew what the most influential customers wanted.

The Palmer Photoplay Corporation and Women

These customers, for better or worse, often appeared to be women. Before the days of more systematic audience surveys, Hollywood lore assumed that women were numerically the most significant audience segment and wielded an authority over the choice of motion picture seen by their dates and families greater even than simple numerical majority would suggest. They were also the cynosure of publishing and advertising. Martin Pumphrey notes that "of the eight American mass-circulation magazines of the 1920s with circulations of over a million, six were women's magazines. By 1928, according to one advertising executive, 97 per cent of all advertising was directed at women and 67 per cent of all consumer items were purchased by women" (184–85). Because of their exposure to film, advertising, and mass publishing, women were undoubtedly of great interest to Palmer, whose ideal student would be caught at the intersection of all three.

The Palmer literature responded with a particularly welcoming tone to women; its house organ was one of the few screenwriting publications that did not oblige women to read themselves into a universal "he" or "man" (that might in fact be exclusively masculine in the minds of some writers). Successful women were included on the board. Lois Weber's *For Husbands Only* was one of the analyzed scripts that came in the Palmer package. Women were well represented as the authors of many of the printed lectures sent to subscribers; they were also prominent in testimonials and in the ranks of the successful alumni/ae, with fourteen of the twenty-five profiled successes in *Little Stories of Success* women.

Clearly, Palmer valued its female customers, as Manker acknowledged. But wanting female students in order to double the pool of prospective students, and valuing the kinds of writing they are presumed to do, are different things. Manker takes a disapproving tone when speaking of the "woman's picture":

> I am not so sure but that the tendency to regard the screen as the voice exclusively of women is largely responsible for this "slump" which we have all been trying to talk down. . . . We have been saying for so long

that our picture theatres are patronized mostly by women, that it is the women who can make or break the success of a production, that, in an abandoned effort to avoid disaster, pictures have been created with the definite aim of pleasing women. The obvious result is the great quantity of scenarios built upon the experiences, preferably the suffering[,] of women. Men, however, do not have the palate for this type of entertainment . . . [and] prefer to stay at home. (Quoted in Kelly, 615)

To add insult to male injury, women's scripts out of Palmer sold in disproportionately high numbers; while women were only forty percent of the student body, they wrote seventy percent of the stories bought for production (Kelly, 615). Although there is no way of checking the accuracy of these figures, venue may explain the disparity between the brochure and Manker's comments; the brochure was designed to lure students to the school, while *Moving Picture World,* where Manker's remarks appeared, was a trade journal aimed at the exhibitor.

How then would it serve the interests of this correspondence school to denigrate the "woman's picture"? Gaylyn Studlar and Staiger have argued that film narratives were a major means of understanding women's place in society since the narratives repeatedly examined the traits of desirable femininity and masculinity (Studlar, "Discourses," 24–25; Staiger, *Films,* 124). Studlar's and Staiger's examples typically have to do with female consumption of male stars (such as Rudolph Valentino) and with narratives (such as Erich von Stroheim's *Foolish Wives*) that explore relations between the sexes in the 1920s. Manker's comments about the prevalence of "women's" stories from Palmer graduates suggests that this cultural debate extended to the sphere of production as well, in which the comparatively high profile of female customers (and possibly mentors) appeared to cause Manker some discomfort on behalf of the industry. While Manker's unease ostensibly arose from declining box-office revenue, some evidence exists that the cultural status of the industry itself was at stake, and that a too-great reliance on female sensibility might jeopardize it further.[21]

We can read Manker's concerns about status in several ways. The first has to do with the social risks associated with the kinds of films that, Manker argued, were turning the American film industry into the "voice exclusively of women"—kinds of films that were alarming to some commentators in the early 1920s, the heyday of what some termed "sex stuff." Women, it would seem, were responsible for the prevalence of "sex stuff" via the plebiscite at the box office, whether or not they actually wrote such pictures. The *Literary Digest* noted that "mothers are hesitating before taking their children to the movies; but mothers who formerly let their children go without guardianship, and even thoughtlessly with themselves, are responsible for their own present consternation" when the children patronize risqué pictures ("Where the Blame Lies," 28). DeMille's sex comedies of the late 1910s and early 1920s, in which a wife retains her husband's affections only by beating the "other woman" at her own game of sexual self-commodification, particularly aroused the anxiety of conservative reformers. Similar stories of sexual self-commodification, in which the heroine "trades up" socially through the production and sale of an attractive body, were regularly offensive to conservative observers of the film industry, as Maltby has argued of the early 1930s ("*Baby Face*," 27–28).

The second respect in which Manker's complaint proves illuminating has less to do with the film industry and more to do with the power of women to declass a profession (such as teaching) they entered in numbers. If the film industry hoped to hold itself out as an authoritative institution, the incursion of numbers of women would not necessarily aid this representation. Accounts of women working in the film industry often (as I have argued elsewhere) took pains to present their subjects as disciplined professionals, working in a male field and having some appropriately masculine characteristics (Morey, "You"). What is perhaps unexpected in Manker's attack is that it came from the fringes of a popular entertainment industry and not from, e.g., a starving modernist renewing Nathaniel Hawthorne's rage over the "damned mob of scribbling women." Nonetheless, it behooved Palmer's spokesman to suggest that the school did not intend willy-nilly to enfranchise all comers and that the traditional cultural hierarchies would be respected, provided that film could enjoy its place in the sun.

Although Andreas Huyssen concerns himself largely with the relations between mass culture and the avant-garde, his exploration of "the notion which gained ground during the nineteenth century that mass culture is somehow associated with wom[e]n while real, authentic culture remains the prerogative of men" seems useful (191). Huyssen argues that, since the 1920s, critics have tried to avoid gendering mass culture as feminine, but that the tendency to do so keeps reappearing (192). Or, as Banta observes of literary production in the age of Taylor, the masculine-feminine question is sometimes hidden, sometimes exposed, in the narrative productions of the culture of management, but it is always there. Males are expected to bring rational practices to an industrialized, mechanized, business-oriented society while females are designated as one major source of that potentially uncontrollable energy surge subsumed under the all-purpose term "the human element." (12)

Applied to Palmer, Banta's remarks suggest that the firm can hold itself out as the mechanism by which the "uncontrollable energy" of female scenarists is properly regulated and ordered; their self-expression is brought within the sphere and tailored to the needs of the film industry. While Palmer appears to offer to enfranchise the unenfranchised (of which women are only one component), the school nonetheless found itself having to maneuver within the surrounding cultural hierarchies. Even so, Huyssen offers a useful counterpoint to the Frankfurt School, with whom we might be tempted to dismiss Palmer students as dupes of clever promotion. While such a view of the marketing strategies of the firm may be warranted, it is perhaps less justified when applied to the consumer's desire to master the necessary and useful content of film narration.

Conclusion

Kathryn Fuller argues that the film industry "was eventually successful in establishing boundaries of fan participation in the creation of films, scripts, and fan magazines. The new limits truncated most avenues of amateur participation in script production and direct lines of contact between fans and film actors," a process she considers to have been under way as early as 1915–17, significantly at the time of the establishment of the classical Hollywood system of narration (116–18). One can argue that the Palmer project was always out of synchronization with the economic and industrial conditions in the industry and was thus doomed to frustrate its adherents. To be sure, exposure to Palmer materials may even have accelerated the shutting down of amateur participation that Fuller observes. In promoting a Taylorized formula for the production of screenplays—indeed, a Taylorization of self-expression—the Palmer Photoplay Corporation may have worked to reconcile students to their fates as members of large industrial enterprises. The audience may even have been urged to consume more films simply because it hoped to have a hand in producing them. The most consistently repeated piece of advice to the would-be scenarist, not only among the Palmer materials but also among trade papers and fan magazines, was to see as many movies as possible, as many times as possible. It is as well to acknowledge that the demonstration of technique to encourage amateur production could also discourage such participation in Hollywood even as it appeared to provoke it, thus relegating the audience to the position simply of better-informed consumer. More specifically, books and articles designed to show what an arduous and dignified craft screenwriting really was elevated the social cachet of filmmaking and filmgoing, but also channeled and controlled discontent over the quality of films and the nature of the film industry. Students were advised not to say that they could make better movies until they themselves had tried.

But in some important respects, the Palmer materials actually could make the audience more authoritative, giving students the kind of social power through knowledge of film they evidently craved. A better-educated audience, one with some mastery of the technology of film production, could be represented as helping to bring about better films through more critical appreciation; in this respect, Palmer can be said to prefigure the strategies of the film appreciation and character education movements (discussed in chapter 5). As Harris observes, the well-informed act of consumption becomes an "assertion of individuality"; "objects have so multiplied [by the 1920s] that the ability to control them, to choose among them, to refuse to be overwhelmed by them, becomes a test of personal strength" (196). Significantly, the desire for inside knowledge about the film industry never disappeared, even though the original conditions for its deployment may have dissipated.

After the early 1920s, however, the same curiosity became a means of controlling the excesses of the film industry. For example, in 1930 the MPPDA sponsored a series of lectures in film criticism, technique, and industry economics at the request of Los Angeles area "previewing" groups, who served, in the days before the Production Code, as an informal censoring body. The film "previewers," mostly women, wanted this information "to make themselves better commentators" (Winter, "Wisdom," 3). Indeed, Alice Ames Winter suggests, the West Coast reviewers possess the knowledge that the Palmer Plan itself offered ("Facts"):

> [The reviewers] have grown wise to technique. They have come to
> know a great deal about this curious, separate world of Hollywood,
> which spreads itself before so many millions of people a day. They know
> something of the mechanics by which effects are produced. They know

the standards to which the producers themselves have agreed; the kind of scenes that ought not to appear.... Previewers are valuable only when they begin to take on something of the character of experts. (3)

While MPPDA sponsorship may call into question the lectures' lack of bias, the women's interest suggests that a desire for informed involvement in Hollywood remained, desire that ultimately took more militant form in the cultural negotiations resulting in the establishment and enforcement of the Production Code.

As the form of Taylorized artistic production *par excellence,* Hollywood is famous for the occlusion of authorship within picture manufacture. Haber notes that "it had been Taylor's express purpose to set up a factory which could run effectively without an outstanding executive on top.... allow[ing] the purposes of the firm and the source of power to appear ambiguous and thus invit[ing] suggestions of expert leadership, industrial democracy, and the ascendancy of a service motive over the profit motive" (166), traits that characterized some aspects of MPPDA leadership of the film industry. In debates about the health of the motion picture industry, a major issue was who was responsible for film content—producer, exhibitor, author, audience, or some combination. The standardization that characterized film production partly explains why authority for film content was not always evident, as does the mystique of the box office, which argued that the audience determined what films got made. Who controls film content and how is the mystery that Palmer offered to reveal to its students; while Palmer suggests that personal benefit will flow from this knowledge, the priests and ministers to be taken up in the next chapter motivate their interest in the film industry through a conception of social or institutional necessity. The case of the Palmer Photoplay Corporation suggests that an understanding of the dispersion of film authorship is at once a method of defining film as a democratic field and a rhetorical trope tending to undermine individual agency, even in a discourse that emphasizes personal achievement. Nonetheless, Palmer (in common with the other case studies examined) emphasizes that some of the functions of author must, in this case through instruction in screenwriting, be distributed to film audiences.

CHAPTER FOUR

"Sermons in Screens"
Denominational Incursions into Hollywood

> For most men, and most circumstances, pleasure—tangible material
> prosperity in this world—is the safest test of virtue. Progress has
> ever been through the pleasures rather than through the extreme
> sharp virtues, and the most virtuous have leaned to excess rather
> than to asceticism. To use a commercial metaphor, competition is so
> keen, and the margin of profits has been cut down so closely that
> virtue cannot afford to throw any *bona fide* chance away, and must
> base her action rather on the actual moneying out of conduct than
> on a flattering prospectus.
>
> —Samuel Butler, *The Way of All Flesh*

At moments of apparent economic and legislative crisis, when box-office
returns dipped and national censorship agitation threatened (as in 1921–22
and 1930–34), the film industry characteristically made a great show of
cleaning its own house. The year 1934 saw the establishment of an enforcement mechanism designed to give teeth to the Production Code of
1930. This unprecedented fierceness is attributable in part to reformist
impatience with the broken promises manifest in an assortment of racy
films produced after the drafting of the Code purported to make presentation of such salacious material impossible. More important, Hollywood found itself having to honor the invitations to public surveillance
and participation that it had so lavishly offered since 1922, at the same
moment that the American public no longer believed quite so whole-
heartedly in the responsibility and competence of American industry.

The enforcement mechanism was the Production Code Administration (PCA), which mandated the submission and vetting of scripts and

final versions of films before the issuance of the MPPDA seal that would allow a work to be played in MPPDA theaters and thus reach a sizable audience.[1] Because Joseph Breen, a Catholic, was appointed head of the PCA in January 1934, and because the PCA was established after Catholic bishops mobilized their flocks to register displeasure at the flouting of the Code between 1930 and early 1934, the enforcement of the Code is typically seen as a Catholic project. So, indeed, is the Code itself, written by Daniel Lord, Jesuit and professor of drama, with the urging and help of Martin Quigley, Catholic publisher of the influential trade paper *The Exhibitor's Herald*. Although the Code enshrined those attitudes toward potentially offensive material that had shaped previous codes of acceptable filmmaking practice, such as the "Thirteen Points" of 1921, the "Formula" of 1924, and the "Don'ts and Be Carefuls" of 1927, some historians identify it as signaling a Catholic capture of Hollywood's storytelling apparatus that would block all serious filmic discussion of social problems in the United States until the 1950s (see, e.g., Black, *Hollywood*, 5).

But Catholics were not the only religious reformers interested in developing an understanding, covering the relations among film production, distribution, and exhibition, that would help to improve Hollywood's product. One thoughtful and well-informed contemporary examination designed to provide the necessary facts to reformers, for instance, was *The Public Relations of the Motion Picture Industry* (1931), published by the Federal Council of the Churches of Christ in America, a Protestant group.

This chapter explores the ways both Catholic and Protestant reformers reshaped and redeployed the rhetorics of a democratically accessible film industry (examined in chapters 2 and 3) to create an environment in which the American public could intervene in the destiny of the industry. While the struggles over the creation of the Production Code have already attracted much critical attention, I propose to link them here with a history of filmmaking by ministers. This approach allows us to read a narrative that is usually presented as a comparatively short, sharp struggle in terms of longer-running and more pervasive anxieties, while simultaneously contemplating a larger range of motivations: not merely a hatred of the licentious, but a ministerial desire to harness the power of filmmaking for religious ends. I also wish to create, over this chapter and the next, a framework for the understanding of both religious and educational incursions into the film industry that presents the film appreciation movement as in effect the heir to earlier, failed Protestant attempts at film control. And finally I propose that what animates both Catholic and Protestant attempts at film industry control is ultimately a form of institutional self-commodification, not unlike the individual self-commodification discussed in chapters 2 and 3, that mandates that religion in some sense create a product that can be sold via film.

If Catholic reformers succeeded in capturing the industry to a greater extent than Protestants, they were at once more authoritarian in bringing direct pressure to bear on producers and audience alike, and more populist in their understanding that film could not afford to be culturally elitist. Thus the rhetorics analyzed in this chapter raise the question of film's accessibility, addressed by both juvenile series fiction and the Palmer Photoplay come-ons, in a new way. Here the issue appears to be industry accountability to the mass rather than to the individual. Social rise and self-promotion are no longer on offer. Rather, the rhetorics culminating in the enforcement of the Production Code must grapple with the revelation that corporations cannot maintain the appearance of serving the public without the public's aid.

Both the Code's secular and religious foes and its proponents agreed on the necessity of justifying the autonomy of motion pictures through an appeal to some form of popular control over their production. Where they differed—at times radically—was in defining what segment of the public should be sovereign and how. Religion thus marks a point at which the democratic rhetorics explored in earlier chapters give way to authoritarian rhetorics that insist upon a hierarchy of taste within American filmmaking and filmgoing. Latent in the religious sensibility of the late 1920s and early 1930s was a concern about the quality of the public's judgments, a conviction that these judgments needed to be guided and controlled through more stringent industry self-regulation (the Catholic view) or through uplift programs (the Protestant view). Like series fiction and correspondence schools, religion constructed film as having power that the active subject might harness to some higher end. Unlike these earlier entities, however, religion frequently focused on the dangers of this power and asserted that unmediated contact with film could weaken the fabric of society. Nonetheless, even the authoritarian hierarchy that the religious discourse proposed had to be described as an outgrowth of some form of popular will.

For a number of Protestant observers of the film industry, the rhetoric of popular sovereignty was primarily to be employed in an effort to turn film into a tool for spreading the Good News. This ambition partook not only of the recognition that film represented a new, efficient way of proselytizing, but also of a sense that film's entertainment value could be deployed to bolster the declining popularity of certain kinds of religious attendance. But Protestant approaches were divided. On the one hand, we note a convergence between Hollywood's rhetoric of sin and redemption and fundamentalist revival; on the other, we see a more liberal expectation of film serving as a forum for the discussion of social problems, including such sexually charged issues as venereal disease, prostitution, and birth control (Staiger, *Women*, 103). The liberal stance taken, for instance, by some members of the New York–based National Board of Review, which became increasingly permissive after 1916 (Fisher, 151), was more and more out of touch, in matters of film regulation and content, with the wishes of the rest of the country. Rural Protestants, as Richard Maltby has argued, were frequently suspicious of the intent and values of the urban and immigrant culture of film production ("Code," 45), to which the National Board of Review, for one, was sympathetic.

The range of Protestant attitudes (from fundamentalist to theologically and socially liberal) appears to have bracketed the Catholic approach, often politically more progressive than Protestant fundamentalism but socially more conservative than Protestant liberalism. The Catholic Church in particular wished to avoid all examination of sexual hygiene, a stand that separated it from the most liberal Protestants but brought it in line with more fundamentalist Protestant groups. Where other representations of sexuality were concerned, however, the Catholic Church diverged from socially conservative Protestants of an evangelical stripe, who were willing to use accounts of sexual sin and redemption to increase church attendance. Robert and Helen Lynd suggest in particular that the advertising rhetoric employed to gather audiences for religious revival meetings was nearly identical to the slogans (otherwise so obnoxious to many religious leaders) that packed audiences into local movie theaters; advertisements for revival meetings, such as "a red-hot arraignment of present-day vices," would do as well as come-ons for salacious films (Lynd and Lynd, 380–81). The Lynds, Kathryn Fuller, and Maltby all note that these narratives of sexual sin and redemption represent an area where religion found itself not only conflicting but also competing with film (Fuller, 96; Maltby, "*King,*" 193–94; Lynd and Lynd, 380–81).

A number of clergymen began to make films themselves in an effort to solve the problem of how the church might appropriately cathect with the film industry, although filmmaking among Catholic clergy was less widespread than among Protestant clergy. The film industry's protestations of its inability to dictate audience preferences tended to call forth demands from Protestants for changes in its economic structure, on the grounds that the economic structures in place prevented community control over film exhibition. Indeed, Janet Staiger suggests that the clerical perception of diminishment of local control was one trigger for church film production ("Conclusions," 355). In contrast to the Protestant focus on economic practices, the Catholic Church was willing, as Francis Couvares points out (142), to accept the business patterns established in the 1910s as the price of some textual regulation. Possibly this willingness resulted from the fact that Catholic priests lacked the history of struggle with the MPPDA that characterized certain Protestant groups' attempts at film manufacture and distribution.

Protestant Disestablishment and Authorship

Ann Douglas argues that nineteenth-century Protestant disestablishment (in which formerly established churches no longer received state funds for their support) obliged clergymen to make common cause with women, similarly economically disestablished through their loss of control over household manufactures. Clergy and women together entered the literary marketplace as producers of sentimental fiction in an effort to recoup the moral authority that their economic disestablishment threatened to diminish. Douglas sees this transformation as marking a moment when religion became exceptionally dependent on the marketplace for support: "Nothing could show better the late nineteenth-century Protestant Church's altered identity as an eager participant in the emerging consumer society than its obsession with popularity and its increasing disregard of intellectual issues" (7). In the keen interest that many religious figures took in film during the 1910s through the 1930s, we may therefore discern an updating of an earlier trend.

To be sure, not all Protestants immediately found the new consumer culture appealing in its effect on such traditions as Sunday church attendance. William Uricchio and Roberta Pearson argue that Vitagraph was one studio that early learned the importance of disarming clerical objections to the Sabbath operation of movie theaters, through the production of religious spectacles such as, in 1909, *The Life of Moses*, which took advantage of an exception granted by New York City blue laws to sacred performances (165). Not only did Vitagraph thus undermine clerical objections to violations of such laws, it was also among the pioneers in the incorporation of clergy as "consultants" in sacerdotal filmmaking (167); this practice would continue well into the studio era with such films as *The King of Kings*, which employed the services of both George Reid Andrews (a Protestant minister affiliated with the Federal Council of Churches of Christ in America) and Daniel Lord.

Vitagraph's experience thus represents an early instance of the collaboration between religion and filmmaking, in which potentially disapproving outsiders became allies. Sacred subject matter was a way of controlling clerical dissent for the film producer, and appeared to disaffected clergy as a superior and more attractive form of proselytizing the American public. Thus, for some commentators, religious motion picture making or exhibition promised a new, powerful, and efficient means of reaching Americans. Richard Alan Nelson reports that Jehovah's Witness Charles Taze Russell was an early proponent of clerical filmmaking, weighing in with the ambitious "eight-hour *Photo-Drama of Creation*," which used both color and sound recorded on phonograph records (232). Clerical filmmaking seems particularly to have suggested itself to those denominations preoccupied by dropping attendance or hoping to win converts. As late as 1929, trade papers reported that Aimee Semple McPherson was planning to incorporate a film company under the name Angelus Productions and had "agreed to appear in [a] talker presenting her life story" ("Aimee McPherson," 32A). Clergy from a variety of established denominations were also eager to capitalize on what they perceived as the innate attractiveness of motion pictures to their audiences, especially children.

Conversely, it was children whom the Catholic Church defined as the audience most vulnerable to the ill effects of motion pictures, a view explicitly stated in the Legion of Decency pledge, which called film "a grave menace to youth" (Black, *Crusade*, 210). Protestant reformers concerned about film's potentially malign power were certainly concerned about children, but they also invoked the adult of "low mental age," a particular problem in the construction of public opinion and for the defense of the plebiscite at the box office. Unlike the backward adult, however, children had the advantage, from the minister's (and industry's) point of view, of being a well-defined and easily identifiable segment of the filmgoing population, as well as one that demonstrably "needed" protection—in some cases *by* films rather than *from* them. Take the 5 March 1924 *Variety* piece on "Methodist-Made Pictures," which notes that Methodists are entering film production with shorts and scenics, and with features as an eventual possibility. According to this commentator, "The interest in pictures shown by the Methodists is reported due to a falling off in church and Sunday school attendance with pictures believed the logical means to secure the return of those who have wandered from the fold" (33). Clearly some of the appeal of filmmaking in the context of religious groups was extrinsic to films' narrative capacities, let alone to the specific religious messages that a given film might carry.[2]

And noting, like the *Variety* article, the desire of the Church to use film to capture the young, Cecil A. Stokes remarks in 1927 that

> the field for experimentation with the moving picture camera within the church community is limitless. Boys and girls, who might otherwise drift away, are ready and eager to fuss and fume with a camera. Women will always act and sew and construct backgrounds. Men will always be willing to occupy themselves with the mechanical processes. . . . The formation of cine clubs for the express purpose of developing religious screen material may do much to further bind parishioners to their religious duties and advisers. (18)

The "something for everyone" rhetoric reveals that filmmaking in some churches was more a social activity than a substitute form of worship. Of course, worship itself struck critics at this juncture as primarily a social activity, so the question became how to make it a more appealing one.[3] If one endorses Douglas's argument that economic disestablishment had the effect of driving its victims to adopt the strategies of the marketplace, it is hardly surprising that threatened institutions found themselves trying to offer an appealing commodity, even if it was not the commodity that their original mission appeared to promote. What linked Protestant reformism to juvenile series fiction and the Palmer

Photoplay Corporation was their joint concern with "getting with the audience" psychologically, illustrated in the series fiction by the numerous depictions of filmgoers' satisfaction with the protagonists' product, in the Palmer literature by testimonials from graduates, and among the ministers by a fascination with attendance figures. Each enterprise was obsessed by sheer numbers of "sales," with quantity assumed to testify to quality as well.

For some moviemaking ministers, in fact, product quality (moral as well as artistic) was beside the point as long as the product could swell the audience. Articles such as Helen Lockwood Coffin's "A Minister as a Movie Maker" (1931) frankly acknowledged the difficulties faced by modern pastors in a world where entertainment was inevitably more appealing than conventional worship. Coffin examines the strategies of the Reverend W. H. Stockton (a Methodist) in Newport Beach, California, where "recreation is the end and purpose of almost every man, woman and child in the place" (8). To combat this sybaritic sensibility from within, Stockton made films of local interest, shooting minor celebrities, night fires at the oil refinery, and boys with their dogs. The primary appeal, quite simply, was to vanity:

> When it is announced that such and such local scenes will be on the screen, every man, woman and child concerned in the action is there to see how he looks. The little church auditorium is crowded. Each meeting has its other features of community singing, friendly intercourse, story telling, games and educational reels. The selection of these gives the pastor many a fine opportunity to put across some needed lesson or suggestion. (9)

Clearly Stockton was no rival to a major studio. His films (and even the more ambitious productions of the Methodists) were inevitably limited in their appeal. Yet, while many pastors acknowledged their dependence on amateur and professional film as the means by which they maintained the size of their congregations, a comfortable modus vivendi between the film industry and religious institutions never arose, because the two were engaged in overt economic competition in the area of theatrical film exhibition. While religious leaders found it hard to trust elements of the motion picture industry not to corrupt America's children, the film industry reproached Protestant ministers with attempting a naked grab for film exhibition profits. It is necessary, therefore, to look briefly at the structural aspects of the film industry that tended to make this competition particularly heated.

Industrial Structures

To begin with, it is important to understand that the MPPDA had lost credibility because of the Depression and not because of any particular financial failure within the film industry. In other words, the Hays Office found itself temporarily discredited because it was an artifact of a particular set of 1920s business methods rejected wholesale by the American public in the 1930s. The rhetoric of public service and business ethics that Morrell Heald suggests justified the significant industrial consolidation of the 1920s also "demonstrated the concern of an increasingly centralized economic system for careful cultivation of its public relations" (Heald, 132). With the arrival of the Depression, however, business self-regulation stood revealed as a bankrupt promise, which encouraged demands for direct government intervention in the structure of the film industry, especially from Protestant groups who felt that their too close association with the Hays Office had compromised their ability to direct the film industry.

As Ellis Hawley notes, "the basic difference between the Hooverian system and that emerging from the New Deal measures of 1933 lay in the latter's acceptance of a larger public sector and of a public obligation to enforce privately developed regulations found to be in the public interest" (*War,* 225). The Production Code manifested this change in public attitude toward industry. While the Code did not have the status of legislation, and thus remained a form of self-regulation, even Catholics opposed to government censorship of motion pictures hoped to see the Code's provisions sutured into the government-sponsored code of fair practice developed for the film industry under the National Recovery Administration, or NRA (Maltby, "Code," 59), a development that would have given the Code the force of federal law. Significantly, as Douglas Gomery argues, the NRA increased the power of the film monopolists at the expense of the small exhibitor and manifested government willingness to endorse the film industry's economic practices (52).

That machinery for the enforcement of the Production Code did not emerge until well into the Depression suggests, as Maltby argues, that the national financial collapse played a significant role in the discrediting of the Hays Office and its former allies ("*King,*" 212). Conventionally, historians have invoked the economic fragility of the studios to explain the gap between the promulgation of the Code in March 1930 and the commencement of more systematic enforcement in January 1934, when Breen, a well-connected man whose Catholicism was inclined to be militant, became head of the Studio Relations Committee, precursor agency to the Production Code Administration (see Jacobs, *Wages,* 109). The latter are variously represented as dependent either upon racy material to maintain box office during the severe business downturn of 1932–33, or upon the good will of bankers. But I contend rather that the revival of Catholic interest in social action, noticed by Couvares as an influence on the formation of the Production Code, may have as much or more explanatory power for the timing of the enforcement of the Production Code.

For the Code is not merely the artifact of a particular economic moment (although it is that); it is also the product of extremely complex cultural and financial negotiations that shaped the relationship between religion and film between the 1910s and the 1930s. Maltby notes that "one 1923 estimate suggested that as many as 15,000 church schools and clubs were showing motion pictures as part of their work. By 1927 the MPPDA estimated that there were 27,000 schools, churches, and other institutions equipped with projection machines" ("*King,*" 190), a comment that hints at the existence of a significant parallel exhibition tier for films of moral uplift and education. Schools and churches not only had their own exhibition resources but also benefited from separate distribution structures, such as the Community Motion Picture Bureau, which flourished from 1913 to 1920 and which sought to assist the local church or municipal group in showing inoffensive commercial films more efficiently (Fuller, 91).

The Hays Office notwithstanding, Hollywood was far from monolithic, and the most significant economic fault line running through the industry separated independent exhibitors from those directly affiliated with the vertically integrated studios that controlled production, distribution, and more closely affiliated segments of the exhibition tier. For a studio-owned theater, production, distribution, and exhibition interests would of course have been identical; such theaters, however, were substantially in the minority. The rest, independent and quasi-independent theaters alike, were not free agents when it came to accepting and reject-ing studio product. Independent theaters often attributed the necessity of playing locally unpalatable films to two mechanisms that the studios had designed to curtail the freedom of the independent exhibitor: block booking, which mandated the rental of a whole menu of films, and blind selling, which required the exhibitor to bid for product sight unseen.[4] Not surprisingly, these mechanisms were the special targets of reform-ers' ire. Clearly, the ongoing tension from this form of exhibitor control concerned centralized challenges to local authority. To the extent that the motion picture industry produced a national product in a uniform way, it undermined traditional notions of local sovereignty.

Because the vertically integrated studios created artificially high pro-duction costs as a deliberate bar to the entry of additional competition, and located most of the economic rewards in the control of exhibition (Maltby, *Entertainment,* 43–48), two possible reformist strategies for di-minishing Hollywood's economic control of the film industry had, from the outset, little chance of success. These strategies were, on the one hand, to produce, distribute, and exhibit morally worthy films in delib-erate competition with Hollywood, and, on the other, to attempt to es-tablish distribution and exhibition networks using Hollywood product, networks that would have the effect of curtailing access to less desirable films by making more moral pictures readily available at the local level. Hope triumphing over experience, reformers promoting film selection programs generally assumed that in the moral economy Gresham's law might operate in reverse: the good film would drive out the bad. It was enough, they hoped, to provide the audience with good material; if the clergy showed it, the audience would come.

Both approaches had basic structural difficulties that worked, by and large, to undermine their large-scale implementation. Even well-produced educational, moral, or Christian films were not necessarily as popular as the "bad" kind—certainly not automatically *more* popular. Maltby for one argues that Cecil B. DeMille's *The King of Kings* (1927) demon-strated that even a religious spectacular made with considerable exper-tise would not necessarily perform adequately at the box office ("*King,*" 208). Even if clergy could manufacture films in such a way as to out-draw Hollywood's product in a fair fight, Hollywood still controlled the distribution and exhibition networks; religious or uplifting filmmaking was chronically undercapitalized and could not pay its way or create

anything like a permanent, autonomous distribution and exhibition network. Moreover, Hollywood objected that, when pastors took advantage
of blue laws to show popular (but presumably appropriate) films on
Sunday, and permitted a captive audience to pay for the privilege of attending with a "silver collection," the ministers were engaging in unfair
competition.[5] Finally, if reformers attempted to work through the distribution and exhibition of Hollywood pictures at venues outside Hollywood's purview (such as churches), they risked alienating the independent exhibitor, their natural ally in any struggle with big business in
Hollywood.

The subject matter of films that religious groups exhibited, like that
of the films they manufactured, ran the gamut from God (sacred subjects)
to Mammon (clean fun). It was the latter that particularly rankled with
commercial exhibitors. Indeed, showings of theatrical films in churches
and town halls were perennially a hot issue in the trade press during the
1920s, flaring up even after the inception of the Hays Office in 1922. In
early 1924, *Variety* reported that "exhibitors in [Utah, Michigan, Ohio,
and Pennsylvania] charge that the opposition from church-exhibitors
has reached the point where it is a question of either stopping furnish-
ing films to the churches or the closing up of their theatres" ("Church-
Exhibited Pictures," 21). Exhibitor anger over such incursions provoked
the Hays Office to urge churches to exhibit only religious pictures.

The mutual hostility between more radicalized (or financially desperate) religious groups and local exhibitors is important because it strength-
ened the Hays Office, whose role as mediator among various segments
of the film industry and outside pressure groups permitted it to play
both ends against the middle. It could claim to be protecting the local
exhibitor from unfair economic competition, at the same time that it
"disciplined" the maverick exhibitor on behalf of local pressure groups.
The Hays Office's apologist Raymond Moley argued, for example, that
the blame for the exhibition of indecent pictures should be placed most
often on irresponsible exhibitors, since "the exhibitor *himself* chose the
sexy picture in preference to the pure one whenever he was offered the
choice" (*Hays Office*, 65).

The independent exhibitor and the reformer nonetheless often made
common cause in opposing centralized control over film exhibition,
and their joint rhetoric in this instance typically emphasized local responsibility to community interests over the soulless standardization of

Hollywood big business. The MPPDA, however, was careful to try to contain the development of reformist opposition to economic strategies, and did much to defuse dissatisfaction with the film industry through outreach to *national* public service groups, which in turn had the responsibility of explicating film industry aims to *local* pressure groups. In essence, the Hays Office hoped to exchange democratic opposition for republican support by creating an organization that would replace direct local pressure for change in the film industry with mediated national representation to and from the Hays Office. The film industry no doubt imagined that a single, centralized public relations entity such as the Hays Office, steeped in the rhetoric of public service, would enable it to control not only exhibitors but also audience pressure groups. Ironically, as Garth Jowett observes, Hollywood's critics had the MPPDA to thank for uniting them: "almost every major national conference on motion pictures was organized by the industry through the Hays Office or the National Board of Review" (*Film,* 200).

To see how the Hays Office's structure made it vulnerable to cooptation from outside, it is worth looking in more detail at how the MPPDA operated. The Committee on Public Relations, an arm of the MPPDA, began operation within months of the latter's establishment in early 1922.[6] A memorandum describing the committee's accomplishments reveals it as a grand republican object in composition and activities, representing (as of March 1923) seventy-eight members of sixty-two national organizations "with an estimated combined membership of 11,000,000," and as having mailed "35,652 letters" and "approximately 385,000 copies of literature." Its aims, according to the same document, were to provide "a channel of communication between the public and the industry, submitting comments, criticisms, and suggestions to the industry, and telling the public of the problems and developments of the industry and about commendable pictures" (Hays Papers, box 19, folder for 20–23 March 1923). In providing many channels for public input into the film industry, Hays hoped to control dissent by appearing responsive to community demand.

At the time of Hays's accession to the role of "Czar of all the Rushes," however, independent film exhibitors accused him of preferring producer interests over theirs; this tension is a concern here since these exhibitors presented themselves as an often frustrated additional channel for the necessary expression of vox populi to the industry. Sydney S.

Cohen, head of the Motion Picture Theatre Owners of America, expressed the view that exhibitors were, in a sense, the directly elected representatives of the people, whose collective power and authority it was their right to wield. In an assertion of the industry's "democratic" underpinnings that recalls the rhetorical positions taken in series fiction and Palmer advertisements, Cohen claimed that

> it became (or becomes) more and more apparent to the man and woman in the theatre management business that their daily contact with the people made them and their theatres public service forces of unlimited potential strength and importance. . . . The theatre owner could not understand the alleged logic which lay back of the socalled trustification idea and permitting elements which had no direct contact with the people whatever to control the Motion Picture business, which in its very nature could exist only through a complete understanding and cooperation with the public.[7] (Speech in typescript, Hays Papers box 16, folder for 14–21 March 1922)

The Hays Office moved quickly to exploit disagreement among reformers and independent exhibitors by representing itself as the one national body capable of uniting all parties concerned about the U.S. film industry. In so doing, of course, it necessarily gave selected industry outsiders the impression that they were now industry insiders. Moreover, the Hays Office actually involved itself in a number of reform projects, projects that spoke to reformers' concern that unmediated contact with film might endanger weaker members of the viewing public.

In playing a game of accommodation, the Hays Office also investigated the possibility of coopting the manufacture of religious pictures, possibly in self-defense against church exhibition of theatrical films. In early 1925, we find another instance of the office's tendency to centralize grassroots activities that might otherwise be a node for local resistance. Lew Wallace Jr. (son of the author of *Ben-Hur*) had the task of overseeing "the development of pedagogical and religious motion pictures" on behalf of the MPPDA (Wallace memo, Hays Papers, box 29, folder for 7–15 January 1926). Evidently William Harmon, president of the Religious Motion Picture Foundation, had promised funding to enable the MPPDA to pursue the possibility of engaging in the manufacture and distribution of religious films. The funding appears never to have materialized (Wallace claimed that much of his work went unremunerated), and Wallace's relationship with Hays's secretary, Courtland Smith, became acrimonious when Smith finally detailed Wallace to MGM to work on the adaptation of *Ben-Hur* there under way (Wallace memo, Hays Papers, box 29, folder for 7–15 January 1926).

Early in his abortive career as MPPDA consultant, however, Wallace concluded that the Hays Office should not merely piece together pedagogical pictures from existing sources, which might suggest a less than firm commitment to the project, but rather should go into new production systematically. In a memo to Smith of 20 March 1925, he argued that

> after compiling such a film [from extant sources] we would still have only a makeshift and not a film with which we could conduct a fair experiment. Frankly and honestly, I can not recommend the adoption of such a course because I believe we would only be making another contribution to the many failures of the past. Unquestionably, a film could be made up from existing material at less expense than the production of a new one, but it would be about the same as making a hoe with which to equip a ditch digger when what we know he really needs is a pickaxe. (Hays Papers, box 26, folder for 20–23 March 1925)

While it seems obvious now that the MPPDA was not likely to fund a major series of religious movies, it was possibly not so obvious to Hays and his associates, because responsiveness to pressure groups motivated this investigation of options. In another 1925 memo, Hays listed Wallace's survey of religious and pedagogical pictures as an example of the kind of information he wanted as proof of MPPDA energy and goodwill, for a major presentation on the organization's accomplishments (11 February 1925 Hays memo to Smith, Hays Papers, box 26, folder for 6–11 February 1925). Significantly, however, the MPPDA's interest in religious motion picture production was short-lived, and appears to have been the offspring of the twin desires to colonize potential competition and to impress pressuring supporters with the organization's zeal.[8]

As the industry sought to expand in the 1920s, it derived some of its inspiration about new markets from its critics. For the Hays Office, then, the development of any sort of parallel film industry was not merely a threat but an opportunity to acquire access to underexploited audiences. At least to some extent, the Hays Office merged its interests with those of its critics in an attempt to increase both goodwill and profits. But ultimately this apparently cozy coalition of film industry and reform groups proved unstable, for two reasons. First, the rift between audience taste and the program of uplift that the Hays Office was undertaking on behalf of pressure groups proved unbridgeable.[9] In other words, two separate plebiscites coexisted, one conducted at the box office (which religious motion pictures typically lost) and the other by committee. Another way of looking at the electoral rhetoric is to suggest that it reflected two separate and opposed voting methods, one democratic (involving direct representation at the box office) and the other republican (inasmuch as it involved representation by committee). As Ruth Vasey notes, "Hollywood's unabated inclination to cater to the wider public's preferences undercut Hays' effort to improve the industry's public relations through the committee" (34–37), a result suggesting just how indispensable the rhetoric of the plebiscite at the box office was. While it was ultimately the republican voting that decided what could be shown, the authority of such a method was always referred back to the public will.

The second reason that Hays's coalition fell apart was that it was really too cozy. The MPPDA, as we have seen, proved exceptionally good at coopting opposition by appearing to accede to external pressure. Perhaps more insidiously, it proved effective at creating outsider participation through payment. Andrews, the Protestant minister remunerated for his assistance on *The King of Kings,* became an implacable foe of the MPPDA in 1927 when further donations, some part of which were supposed to come from the profits of *The King of Kings,* were not forthcoming to his Church and Drama Association (Maltby, "*King,*" 207–8; Couvares, 136). Not only does Andrews's involvement in this film, along with his dashed hopes of sustained and profitable religious film production under his control, illustrate the potential for rivalry between activist ministerial filmmakers and Hollywood, but his receipt of funds from the film industry suggests why the appearance of greater public involvement was made so necessary by the Depression. The rank and file of major social and religious organizations may have felt that their leadership had been bought off and was consequently abandoning its mission.

The apparent corruption of reform leadership continued to be an issue into the year 1930, when thirteen civic organizations felt it necessary to issue a denial of "receipt of Hays money" ("Civic Leaders Deny Receipt of Hays Money," 22). Indeed, the Federal Council of the Churches of Christ in America found itself having to audit the activities of its own general secretary with regard to the receipt of "honoraria" from the Hays Office (Shenton, 113–14). A prominent Catholic member of the ecumenical group of organizations issuing the denial, Mrs. Thomas A. McGoldrick, capitalized on this moment of Protestant confusion by noting that

> there is no group in a better strategic position to answer such "charges" than the International Federation of Catholic Alumnae. . . . [who have] worked intimately with Mr. Hays for eight years. . . . In these years we have never received one dollar of remuneration from the motion picture industry or the Hays office, and what is more important still, we have never been offered a penny from anyone connected with motion pictures. ("Bold Racketeering," 22)

McGoldrick also observes that "it is significant that some of these charges are being brought by men who have actually demanded participation in gross receipts of pictures for the alleged services they render," which may be a dig at Andrews ("Bold Racketeering," 22). Protestant leadership had failed in its chief aim, disinterested public service. Not only had the film industry not been cleaned up, but it had seduced some of its severest critics, perforce no longer in touch with the public will because corporate special interests had so successfully absorbed their leaders.

The Construction of Catholic Authority

Into the vacuum left by the disappointing performance of Protestant reform groups stepped a newly vigorous and well-organized Catholicism. Reminding us anew of the importance of religion to commentary on Hollywood, at least one Catholic observer chose to promote on sectarian grounds the Church's right to guide the film industry by identifying the moral failure of Jews and "pagans" to serve the public interest. John J. Cantwell, Catholic archbishop of Los Angeles and San Diego, looked at Hollywood at length in an *Ecclesiastical Review* article in February 1934, a moment when the film industry was between the economic rock bottom and the revival in profits about to begin; for Cantwell, the financial difficulties of Jewish studio heads such as William Fox and Adolph Zukor offered cause to hope that Christian direction of investment in the film industry would improve its quality. The timing of this article is also significant in helping mark the inception of the Legion of Decency campaign, by articulating the concern expressed by the group of Catholic bishops who determined in November 1933 that enforcement of the Production Code was needed. The Legion campaign proper emerged in the spring and summer of 1934.

Cantwell suggested that the financial leadership in the film industry was on the verge of major change because of the bankruptcies of Paramount and Fox and the general debility of the remaining studios; it would, he predicted, pass into the hands of brokers and large corporations such as American Telephone and Telegraph (AT&T) (Cantwell, 138–39). But he was concerned that this change would not be enough to secure the safety of the nation's youth. Rejecting the excuse that it was necessary to play to the lowest common denominator to keep business during the Depression, Cantwell argued that the blame for "evil" films rested squarely on the shoulders of studio personnel (of whatever ethnicity) who were not Christian:

> Now, it may be asked, who is responsible for all this vileness and worse, that is being made to serve as the instrument of debauchery of the youth of the land. The Jews? Yes—and no. Most of the producing-distributing-exhibiting companies are operated and managed, when they are not actually owned, by Jews. . . . Jewish executives are the responsible men in ninety per cent of all the Hollywood studios and it is these Jewish executives who have the final word on all scenarios before production is actually launched. Certain it is that if these Jewish executives had any desire to keep the screen free from offensiveness they could do so. (142–43)

But where Jews were going especially wrong, Cantwell seemed to suggest, was in their terrible choice of employees, particularly screenwriters, who included "hundreds of playwrights from the Broadway stage" and "writers of the pornographic school whose books have had a great sale in recent years. It is from these men and women that the stories now current on the screen are selected. *Seventy-five per cent of these authors are pagans*" (143).

Anti-Semitic rhetoric directed at film personnel was hardly a new phenomenon in the 1930s, a decade in which many commentators questioned whether America could afford to permit these perennial outsiders a free hand in the shaping of its most important mass culture industry.[10] But Cantwell's rhetoric is significant to the present chapter because it suggests the kind of conceptual rearrangement that would be a prerequisite for the enforcement of the Production Code. Not only did Cantwell reject the box-office rhetoric tacitly endorsed by the liberal National

Board of Review; he also argued that the people charged with the direction of the film industry did not represent America, but were Jews or New Yorkers or pagans (and very likely all three). Cantwell suggested that while "It is true *in a measure* that this sort of stuff seems to be the stuff that American audiences want" (144; my italics), nonetheless those who chose and produced American film subjects were members of a minority ready to foist its tastes on the rest of the nation. He observed that

> in any discussion having to do with moral values in motion-picture
> entertainment it is the practice with many of the producing people in
> Hollywood to point to the box-office success of pictures, which to many
> patrons, are definitely offensive. As a matter of fact, the records indicate
> that few pictures that are vile, suggestive or unclean attain to any
> outstanding success. (145)

Cantwell's rhetoric has important ramifications. It suggests that the Hays Office was out of touch with the wishes of the public because it was never directly in touch with the people. The leadership of liberal Protestant groups, one might argue, was also unclear about the popular will, a point that Couvares makes of the largely Protestant but ecumenical National Board of Review, which was ultimately unable to unite the various Protestant attitudes toward motion pictures, whether across denominational lines or, more fatally, across the urban/rural divide (Couvares, 130–31). Insofar as by the 1930s the American Catholic Church was a more urban than rural institution, its constituency was quite desirable in the eyes of the MPPDA.[11] Urban box office was the most profitable sector of the film industry. The MPPDA clearly desired the good will of the single largest American denomination, concentrated not so incidentally in the communities already most important to the MPPDA's own business strategies. The irony of this confluence of Catholic concern and film industry structure is that the plebiscite at the box office remained important, Bishop Cantwell notwithstanding. The rhetoric suggests that, while methods of franchise other than paid admissions were coming to the fore in measuring the film industry's responsiveness to both moral arbiters and the public, audience size was clearly still crucial.

In fact, control of audiences was the very essence of the Legion of Decency campaign, which in its more extreme manifestations asked Catholics to refrain not only from attending immoral movies but also from patronizing good films at theaters that sometimes showed bad ones. The Legion of Decency, while Catholic in philosophy and impetus, looked at least initially like an ecumenical project. Moley notes that New York and Philadelphia rabbis endorsed its program and that Missouri Lutherans likewise found it appealing (*Hays Office*, 81). Ecumenism was otherwise discernible in the Legion's reliance upon such standbys of the "taste-improving" liberal Protestants as rating films to suggest suitability for a particular audience. The Legion's pronouncements, however, were not merely guides to conduct; they were unappealable fiats backed by the full authority of the pulpit and the episcopate. To take the pledge required for membership in the Legion was to accept not the sovereignty of one's own taste and conscience but rather the authority of the raters.[12] To attend a film to which the Legion objected could be a serious act of disobedience, at the very least.[13] Clearly the Legion's approach, for all satisfied with it, spelled the end of the project of improving audience taste. The Legion's project appeared analogous to the production of an Index for film.[14] A bad film was so toxic to the soul that priestly mediation between the audience and the film industry was needed.

What a priest would do for an individual member of the flock, the Catholic Church undertook to do for the nation as a whole, a response pragmatically tailored to the nature of film as an industrial art form by addressing the contagion at its source. Father Lord's attitudes toward other damaging artworks made clear the important role played by the priest (especially vis-à-vis children) in ensuring that mainstream cultural mores did not undo the virtues of a Catholic education. Lord is significant not only as the author of the justification for the Production Code but also as one of the first Catholics to institute a program of blacklisting, which he did in his magazine *The Queen's Work* (Walsh, 97). A tireless producer of pamphlets on many subjects, he took up the problem of the young Catholic's desire to stay current with the exciting world of modern literature. In his *I Can Read Anything* he creates a dialogue between sympathetic Father Hall and a brother and sister who find themselves beginning to experiment with racy reading. Lord's approach, through the persona of the priest (who, like Lord, dabbles in various forms of writing) is to demolish one by one the children's objections: nothing one reads can hurt one (an act of faith should be enough to repel the mental invaders, suggests a child, apparently erroneously);

one needs to keep current with developments in literature (not this way, counters the priest); one needs to know about the seamy side of life for one's own protection (again, says the priest, not this way); nobody really takes these literary attacks on virtue seriously (not so, says the priest); and so on.

The children are initially certain that nothing can shake their faith in God and their Church, and that they thus risk nothing in exposing themselves to modern literature. Here they are as naive, counters Father Hall, as those hapless and similarly confident Protestants:

> You're not too young to have heard of the systematic attack on religion called Modernism. Modernism set out to blast the whole ground from under faith, destroy the supernatural, and reduce to the level of folklore all religious belief; it meant to take God out of religion and to substitute on His altars humanity. In no time Modernism had wrecked whole sections of the Protestant churches and turned them from faiths into social agencies. (18)

Lord's diatribe against modernism in this pamphlet suggests that more than sex was at stake in the Production Code, although the popular perception often reduced its precepts to prohibitions against lewdness in film. While the studios were not fighting over the rights to *The Golden Bough* and William James's pragmatist *The Will to Believe* (two works that come in for special condemnation as representatives of modernism), other more accessible works might carry the same taint.

The hope of liberal Protestant groups, such as the National Board of Review, that film might be capable of the same artistic experimentation that was going on in other art forms was an alarming concept to anyone who believed that modernism was the doorway to loss of faith. Liberal Protestants were less interested in controlling the public's access to potentially dangerous content than in providing viewers with the weapons to combat unhealthy blandishments from the media, a stance that appeared naive to Catholics such as Lord's Father Hall. Indeed, the whole pragmatist project of investigating received truths was horrifying to some Catholics. As William Halsey puts it, "the common enemy [of some Catholic thinkers] was the instrumentalist and empirical thought that rejected a purposive and rationally predictable cosmos in favor of a relative and contingent structure of reality" (4). It was not enough to be charitably interested in the welfare of one's fellows; one had to believe in God as well, and anything that threatened that primary attachment was anathema. Francis Augustine Walsh consequently dismissed pragmatism as a manifestation of inappropriate affect, "the canonization of thrill, the scientific delineation of the age-old rejection of the customary and usual for the advanced and different" (quoted in Halsey, 138).

Following from the rejection of pragmatism and its works, therefore, was a Catholic disdain for discovering audience taste. No amount of popularity would excuse a preference for error, so the Catholic reformers were under no compulsion to determine what audiences actually liked if their choices could be construed as harmful to them. Maltby observes that the Production Code substituted for audience research ("Code," 40) and thus sidestepped the question of audience likes and dislikes altogether in the interests of stating uniform standards, a corollary that follows from Catholic insistence upon standards derived from "natural law." This emphasis on natural law was undergirded by the Thomist revival within Catholic philosophy. Neo-Thomism was a deliberate renascence of the scholastic philosophy of St. Thomas Aquinas, and promoted the idea of a universe in which humanity could more nearly approach its creator through the exercise of reason.

While some producers might have appreciated the apparent simplification that the Production Code thus represented, of the always onerous task of predicting and meeting audience taste, the will to understand the nature of audience desire did not disappear altogether and certainly would have been required by any attempt to serve a segmented rather than unified constituency. Reformers associated with the educational establishment, very much under the influence of "group process psychology" and the intellectual offshoots of pragmatism (see chapter 5), kept this approach in the foreground.

In addition to the emphasis upon "natural law," and to some extent following from it, social activism was another component of the new authority with which Catholics were entering American life. Natural law allowed Catholics to appear as defenders of America's founding ideals, which were predicated on similar philosophical bases (Couvares, 150–51; Halsey, 4–5). In part, the newly prominent Catholic role in American politics, illustrated by events such as Al Smith's 1928 presidential bid, resulted from what Richard Hofstadter identifies as the "pride and self-consciousness" of the now assimilated "descendants of the great wave of the [immigrants] of the late nineteenth century" (298). As Hofstadter points out, some Catholics' readiness to involve themselves in

public policymaking stemmed from a rejection of 1920s nativist movements such as the recrudescence of the Ku Klux Klan (299). The popular demagogue Father Coughlin, for example, began his broadcasting
career in 1926, ostensibly to scotch the power of the Ku Klux Klan in his
Detroit suburb, Royal Oak (Brinkley, 82).

While Couvares argues that the campaign for clean pictures demonstrated that, among other things, Catholics wanted to take action to
prove themselves responsible citizens concerned about the public welfare (150), activist Catholics may have hoped to achieve more than the
admiration of the less-organized Protestants. In them burned an ambition to raise the cultural profile of Catholicism, as Halsey remarks:
"ready to take advantage of this moment [of disaffection with materialism, in the 1920s] was a highly self-conscious group of Catholics who
believed that the 'Catholicization of America was very possible'" (143).
The combination of accessible philosophy and a sense of social mission
gave Catholicism extraordinary social energy in the 1920s and 1930s,
and clearly Coughlin, for one, exercised real political clout; when asked,
radio audiences said they preferred his broadcasts to philharmonic
concerts, and when listeners were asked to oppose Senate ratification of
Roosevelt's 1935 decision to enroll the United States in the World Court,
Coughlin's followers expressed their sentiments to their representatives
and utterly ended the project (Brinkley, 120 and 135–36).

Hays made a trip to Detroit to meet with Coughlin following a ru-
mor that the latter was about to join in the denunciations of the film industry in the summer of 1934 (Hays, 462); however, I do not mean to
suggest that Coughlin personally endorsed the Legion of Decency campaign.[15] Rather, I propose that we read Coughlin's authority as part of
the larger context for public agitation against the film industry during
the Depression, since we may discern striking parallels between the structure and organization of his ecumenical crusade for "social justice" and
the Legion of Decency's campaigns. Both represent a populist revolt
against the 1920s business methods dear to the Protestant establishment. If, as Hawley argues, the MPPDA was one of a number of trade
groups manifesting Hooverian associationalism, this was exactly what
made it so liable to rejection once the Depression made itself felt ("Facets,"
96). An unresponsive film industry looked more like an unwelcome
"special interest" during the Depression than before, and was thus open
to the Legion of Decency's variety of populist censure. Thus, it is not

necessary to argue, as Stephen Vaughn does, that financiers close to the Catholic hierarchy pressed individual studios into adopting the Production Code during the industry downturn of 1933; in fact, the most sustained pressure to conform to the Production Code came in 1934, the year box-office figures began to recover from their nadir (Balio, 30).[16] The studios' reputations as socially responsible corporations, rather than their actual fiscal health, predisposed the Hays Office to respond to Catholic demands during the summer and fall of 1934.

The similarities between Coughlin and the Legion of Decency started with a distrust of national centers of power and an appeal to a large audience across sectarian lines at the expense of specific outsiders. In Coughlin's case, the targets (which found endorsement among Americans of several faiths) included Wall Street, Washington, and, beginning in 1938, Jews (Brinkley, 269). As Cantwell's remarks about Jewish responsibility for films' "vileness and worse" help to illustrate, anti-Semitism was not unknown among the leaders of the Legion of Decency campaigns, and Cantwell's ghostwriter Breen, enforcer of the Production Code Administration, was notoriously anti-Semitic in his remarks about, and to, Jewish studio heads (Walsh, 84). Although Breen was a studio insider, he believed that the Legion was a necessary tool for the enforcement of Catholic aims, and he was initially willing to provide information to strengthen the position of Catholics outside the industry. That Breen was a Catholic also helped to tame and localize the foreign body that was Hollywood for its Catholic critics (Walsh, 83), even as his position gave them greater national authority. Coughlin, too, suggested that he was the people's representative in the corridors of power in Washington, and some of his admirers took him for a beneficent government insider.

Ultimately, then, both Coughlin and the Legion of Decency depended on grassroots response for their authority. In the Legion's case, one might crassly suggest that the box office of organized religion was vying with the box office of Hollywood. The Legion succeeded because it orchestrated an outpouring of popular feeling (not solely Catholic feeling) about motion pictures. As with Coughlin, the articulation of the problem came from above, and the response from below was an apparent eagerness to oppose centralized powers that threatened local autonomy.

Both Coughlin and the Legion gave the appearance of greater national organization than they actually possessed, however—an aspect of their campaigns that hindered consistent and unified action. The grassroots appeal in each instance engendered a certain desire for local self-determination—what Paul Facey refers to as the Legion's "decentralized structure" (57)—although the Depression created the economic conditions that permitted more powerful prelates, such as Cardinal Mundelein of Chicago, to consolidate control over individual dioceses by attempting to standardize, for example, certain aspects of the Catholic relief effort (Cohen, 221).[17] Nonetheless, as Walsh notes of the Legion of Decency, individual priests and dioceses were initially given wide discretion, and one aspect of the campaign that caused considerable anxiety for the Catholic insiders in Hollywood (such as Breen and Quigley) was that individual expressions of resistance to Hollywood were not predictably subject to control from above.[18] Prelates such as Cardinal Dougherty—who called for all Philadelphia Catholics to avoid the movies, a ban evidently still in effect at his death in 1951 (Walsh, 101–2)—could act unilaterally in ways that sponsors of the Legion's aims found threatening because extreme. In Coughlin's case, his National Union for Social Justice drew upon a network of sentiment created by his radio broadcasts, but Coughlin did not promote a systematic or national organization because he was unwilling to delegate authority (Brinkley, 179); as a consequence, his followers at the local level were more interested in providing local solutions to local problems than in mounting sustained opposition to extant national structures, a failing that also dogged the administration of the Legion.

Both Coughlin and the Legion found it effective to combine the imagery of the plebiscite—the unanimous endorsement of the radio audience or the congregation of right-thinking individuals—with direct pressure on representatives through letter-writing campaigns. If Coughlin proved unable to create a national organization (Brinkley, 175–79), the Legion and its sympathizers were far better organized, if more limited in scope, because they had access to preexisting machinery (e.g., the diocesan newsletter and the hierarchy of Catholic educational institutions), which they were willing to use as important adjuncts to exhortations in sermons.[19] Young people in particular found their teachers urging them to write to studios and to stars to register disgust with the prevailing product, just as Coughlin urged his listeners to address directly their representatives and the White House on issue after issue.

Clearly, the appeal of both campaigns was that they fostered accountability to the public, the very thing that the Depression revealed to be lacking in governmental and quasi-governmental structures. Although the Legion of Decency effort was not thematically connected to Coughlin's political program, the two were contemporaneous outpourings of the same populist frustration at conventional leadership. They must also be read as one possible realization of Lippmann's call for methods of audit, although not necessarily in the form for which Lippmann called. Coughlin was a master of the sort of narrative manipulation that disturbed Lippmann, and while Coughlin suggested that he had the necessary economic expertise to address national problems, this was not the case; in effect, he showed that the manipulation of affect was an acceptable substitute for information and debate. The Legion, too, proceeded by methods more passionate and wholesale than scientific, and consequently perennially had to face the problems arising from promulgating subjective judgments with excessive authority.

Hays's meeting with Coughlin deserves some attention in this connection, because it suggests that one facet of this populist revolt was the sort of struggle over authorship and authority with which this chapter began. If nineteenth-century Protestant ministers hoped to recoup lost authority through incursions into the marketplace, beginning in the 1830s, only to find that becoming a commodity undercut the force of the original message, as Douglas suggests, what did Catholic priests hope to accomplish, and what did they risk, through similar incursions a century later? Hays's meeting with Coughlin suggests that some Catholics, Coughlin among them, were willing to stake their prestige in a bid for acknowledgment of an expertise in public relations and even authorship that they may or may not have had. While Hays is extremely circumspect in his memoirs about what he and Coughlin discussed during his visit in August 1934, he notes Coughlin's warmth toward the new media, as Coughlin averred that he "by no means 'had it in' for [movies], and that he felt that the producers, although they made mistakes, had made great headway. He thought motion pictures had the greatest popular influence of any medium of expression—perhaps even greater than the churches!" (Hays, 464). What Hays noted in an unpublished memo, however, suggests that Coughlin had an extraordinarily self-satisfied view of his own powers as a public figure (and in 1934 who would have gainsaid him?) and took Hays's unannounced visit as one made hat in hand to a more knowledgeable figure.

Hays recorded his impressions of Coughlin in this memo in such a way as to suggest that Coughlin was caught up in the problem of catering to his public, in terms strikingly similar to those used in defense of the excesses of the box office. Hays claimed that Coughlin's

> purpose was to find out through an elaborate scouting and inquiry system exactly what would "go over" with the people. In fact, he said that he had that very kind of machine ... that he only planned to talk about things that would appeal; that he was engaged in promoting his following by seeking an issue that would appeal to them. ("Memorandum in re: Visit with Father Coughlin," 6, Hays Papers, box 47, folder for December 16–31, 1934)

Coughlin offered Hays the kind of solution to Hollywood's public relations problems that Hollywood itself had pioneered. On the subject of stars who might appear out of line, Coughlin suggested that Hays "get after the bad publicity given the stars; publicize the good that motion pictures do," a comment that seems strangely unaware of just how expert Hays was in this department ("Memorandum," 2). Coughlin proposed another well-worn Hollywood method—that is, disciplining the comparatively less powerful members of the studio community, the talent, in the interests of good publicity for their masters. For the actors, he suggested a school (at which he would happily teach) designed to "develop their knowledge of public obligation, their cultural values, their capacity to occupy the high place in public influence they are brought into" ("Memorandum," 2–3).

Without a doubt, Coughlin was an unusually media-savvy man. It would be perilous to extrapolate from his experience and temperament to the attitudes of other clergy, who often disapproved of him and his program. But Coughlin's unexpected condescension to Hays suggests the stakes of the battle going on here. To use Couvares's term, it is a "Kulturkampf" or struggle over cultural authority (129), in which the victory goes to those who manifest the best understanding of what the public wants and of how to provide it.

What the Legion of Decency offered its adherents was a say in the running of the film industry; this offer employed the same rhetoric (in a different form) that the film industry and its satellites, such as the Palmer Photoplay Corporation, had employed earlier; if one could not sell a scenario to a studio, one could still, as an individual filmgoer, dictate the terms on which other people could produce scenarios, either by writing to producers or by pledging not to attend certain classes of film. If the prospect of becoming a star seemed more remote in the straitened national circumstances of the Depression, ordinary citizens could nonetheless force stars to toe the line through individual letter-writing campaigns urging them to avoid roles in salacious films. If Palmer Photoplay offered participation to the American public in an era of rising personal and economic expectations, the populist revolt presided over by the Legion of Decency was a similar proffering of access to the film industry tailored to a moment of declining expectations. If the literature of Palmer Photoplay in the 1920s figures consumption as production, then nonconsumption or selective consumption is the inverse of that rhetoric (but to the same effect) for the Depression. The major distinction between these two manipulations of consumption is that Palmer offered success to individuals ready to accommodate themselves to an admired industry, while the Legion suggested that the public was powerful when organized to take control of the industry.

In some respects, this populist rhetoric traces its lineage back to Progressive Era ideologues, who were preoccupied with the relationship between corporate "special interests" and the "people." Richard Pells observes that when John Dewey commenced writing about the national crisis unfolding in late 1929, he continued to use the rhetoric and oppositions familiar to him and his audience from decades before (51). Pells's analysis of the attitudes of Depression Era writers, from anthropologist Ruth Benedict to educator George Counts, suggests the difficulties in attempting to differentiate Progressive Era holdovers from the new thinking engendered by the Depression. For one thing, the Depression gave new urgency to examinations of the place of the individual in a liberal democracy, an issue that had exercised prescient Progressives such as Dewey for decades. If there is a discernible change in these examinations, it appears to be a shift in emphasis from a notion of the completely sovereign individual to a conception of the necessity of individual accommodation to the larger society. Counts, for example, asserted that it was the mission of education to promote cooperation as essential to democracy, saying that the "welfare of the ordinary individual can be advanced only by making paramount the general good" (quoted in Pells, 115).

This distinction sheds light on the present chapter. The Depression-era downplaying of economic individualism suggests that filmmakers can no longer disregard the "general good" in the interest of box-office success; this rhetoric clearly chimes with Progressive Era thinking about the primacy of the people relative to "special interests." Yet the Depression revealed that the New Era solution of professional leadership, such as that offered by "disinterested" reformers, was similarly bankrupt; by 1930 it was clear that the Protestant reformers had not been adequately "disinterested." Despair over competent public opinion in the 1920s notwithstanding, the 1930s mark a moment when the mass is the touchstone of what is "democratic," and even the successful individual must be subordinate to it. True public participation, as opposed to individual social rise, is the ultimate value.

The Hays Office itself reinforced the impression of successful public participation in the Legion of Decency campaign. In considering the Legion's triumph from the vantage of 1945, Moley concluded that the project of improving picture quality was so long delayed because the electorate was disorganized and thus mute: "Only the appearance of some such spectacular guarantor of mass support as the Legion of Decency could lessen the risks of such a venture. Only when support was mobilized for all to see, could Hays bring to swift fruition the work of the previous twelve years" (*Hays Office*, 85). And for any Protestant (or moral relativist) who might suggest that people are not genuinely active participants when told what to think, T. O'R. Boyle, writing in *Commonweal*, suggested that the Legion's campaign worked precisely through its appeal to the individual conscience:

> The natural law, which is but an expression of right reason, a judgment of our conscience, tells us that we may not read a book that is dangerous to our faith or morals. The same law applies to moving pictures. Our censor is our conscience, and this is the censorship which the Legion of Decency demands and is fighting for. The curtailment of the manufacture of immoral films is desirable (and here the action of the Legion is meeting with success), but secondary. (227)

Significantly, the title of Boyle's article is "We Are All Censors," emphasizing the local, personal, and participatory aspects of the Legion's program while sidestepping its authoritarian nature.

As Maltby and Couvares argue, the Catholic Church was ideally structured for cooperation with Hollywood, for reasons ranging from the

Church's emphasis upon natural law to its apparent unanimity as a denomination and its organized hierarchy (possibly more apparent than real [Maltby, "*King*," 212; Couvares, 149]). If we ask what this cooperation offered Hollywood, we find that the Church's *authority* was as valuable as anything. One may thus read the Production Code as the end point in a series of efforts to peg the value of Hollywood's product to an extrinsic authority, ranging from Adolph Zukor's Famous Players Film Company, which affiliated film with "a theatrical model of 'fine acting'" (deCordova, *Personalities*, 45), to Samuel Goldwyn's Eminent Authors program in 1919. Even the star system may look like an external authority, because commentators represented it as an adjunct to the plebiscite at the box office; as Louis B. Mayer remarked, "the public makes the stars." Indeed, the long debate over the power of popular demand to shape film texts was simply the most significant manifestation of this need to find an authority outside the film industry to ratify the products of the industry. With the 1920s collapse of faith in the intelligence and self-restraint of the audience, however, Hollywood had to confront the necessity of identifying yet another patron for its productions. The great virtue of the intervention of the Catholic Church was that it provided an explicit structure for the creation of texts without vitiating Hollywood's ability to locate the source of that authority, again, in the audience. And the Church was gratified by the manifestation of its power, expressed in popular terms. As one enthusiastic observer noted of the early successes of the Legion's campaign, "the Catholic church . . . could put anything through it wished, and could crush anything" (quoted in Walsh, 111).

This last comment suggests that authority is a vexed issue, with many ramifications; Hollywood was clearly not the only institution hoping to locate its raison d'être in the popular will in the 1930s. Given the issue with which this chapter opened, the need of mainline Protestant churches in particular to find a more efficient and more forceful means of reaching a public that appeared to be falling away, it would seem that to stake one's prestige on popularity could be a risky business. Indeed, the later history of the Legion of Decency suggests that as mores changed in the wake of World War II, the Legion risked the prestige of the entire Church in campaign after unsuccessful campaign, to the frustration of Catholics who believed that the ecclesiastical hierarchy should not be yoking the majesty of the Church to the comparatively trivial issue of film attendance (Walsh, 253–55; Black, *Crusade*, 144).

Motion pictures attracted the attention of clergy not just because of films' potential for evil, but because of their apparent potential for good. Even here the Catholic Church may have considered itself especially well placed to take advantage of the new medium, because of its presumed greater readiness to teach spiritual lessons through images. A Protestant minister acknowledges that the popularity of film seems to require religious teaching to take a new tack:

> There are two methods of reaching the mind with truth: through the eye and through the ear. The church has confined itself almost wholly to the ear. With the coming of the motion picture the discovery was made that people could be taught the most important lessons through the eye. I have seen pictures bring greater spiritual blessing than the sermon. I do not see how any active church can do without them. (Quoted in Stokes, 19)

The problem with granting the power of motion pictures as a medium of religious instruction is that this conclusion obliges clergy to attempt to control film in some way, if only to reap its advantages and to limit the dissemination of error.

Commonweal takes the occasion of the release of *The King of Kings* to suggest that the Protestants have an unduly naive approach to the representation of Christ in film, perhaps from inadequate experience with the aesthetic aspects of representation within religious discourse: it is the Protestants who demand tawdry, conventional, predictable religious imagery, and they do so because they are starved for imagery. As R. Dana Skinner explains,

> There are people—thousands of them—whose form of religious worship supplies them with none of the pictorial reminders which human nature seems to crave. The Catholic, accustomed as he is to the Crucifix, the Stations of the Cross and the paintings which for generations have sprung from the devotion of the greatest artists, cannot well escape a constant and vivid mental picture of the Passion.... To the Catholic, then, whose power of forming mental images has received this constant training...the concrete detail of a presentation such as *The King of Kings* [italics added] will do violence to a much more vivid and sacred mental image already existing. (14)

This review suggests that a certain nondenominational struggle over who controls the power latent in the film medium—God or Mammon, broadly—might go hand in hand with conflict over denominational issues.

Significantly, Hollywood insiders such as Breen and Quigley effectively crushed any later tendency to use the screen to preach a particular course of action (except in those general terms enshrined in "natural law"), by relentlessly emphasizing the screen's role as entertainment, not pulpit, and above all not campaign platform. Here we find one aspect of Catholic action undercutting the larger liberal project of Catholic Action; the tradeoff in the Production Code was clearly to be no sex but no politics either. This compromise arises partly from the suturing of Lord's philosophical presentation of the reasons for a Production Code to Hollywood's traditional lore about not offending pressure groups important to the success of a picture. Indeed, Irish Catholics were early prominent among these constituencies, as Couvares's and Walsh's accounts of the reception of *The Callahans and the Murphys* (1927), a film playing on rather hackneyed ethnic stereotypes, make clear (Couvares, 145–47; Walsh, 102).

That the Catholic attitude is clearly the one that the Production Code enshrines suggests that we should award the palm in the struggle over authorship to the Church. After all, we encounter, as Les and Barbara Keyser demonstrate, image after image of the "fighting priest" in such films of the late 1930s as *Angels with Dirty Faces* and *Boys Town*. I have argued elsewhere that the priests in these two films are notable for their ability to rewrite the realist text of the ghetto and crime reporting (Morey, "Priests," 10). In *Angels with Dirty Faces* (Michael Curtiz, 1938), Father Connolly (played by Pat O'Brien) is so powerful an author that he is able, either through personal suasion or through prayer, to cause his childhood pal Rocky to "die yellow" in front of the press. This cowardly death is essential if Connolly is to save the ghetto children from a similar fate, given their fervent desire to emulate Rocky's gangster glamour. The film suggests that the press is not capable of knowing, let alone reporting, the significant news of Rocky's salvation. The film viewer is privileged to know how the Church accomplishes these miracles, but the hard-bitten crime reporter is not. *Boys Town* (Norman Taurog, 1938) takes a similar approach to the contest between newspaper and priest, in which Spencer Tracy's Father Flanagan is shown capable of "rewriting" the boys' lives by providing them with a better environment. In so doing, he frustrates the narrative expectations of the local newspaper editor and again demonstrates the ability of the Church to shape stories to its own ends.

As we have seen from the previous two chapters, the sale of affect was a major lure of the movies. Telling a personal story is the means of entry into the film industry for both the juvenile stars of series fiction about the industry and those who subscribed to the Palmer Photoplay Corporation's come-ons. In both instances, the control of affect is the means of acquiring greater social power. The Catholic Church clearly agreed with this rhetoric about the film industry; affect is what is for sale in film narratives, and it is important that this affect be properly edifying. It matters who controls the construction of affect, and it would be irresponsible to leave such production to the tabloid newspapers or to a completely unregulated film industry. In casting aspersions upon the ability of Protestants to produce acceptably edifying affect through visual means, the Catholic Church is merely reasserting its greater experience with the rich, multimedia productions that gave it such an affinity for film. While, as we shall see in the next chapter, reformers of a more technocratic stripe may have distrusted affect, the Church emphasized that affect had its place, under proper supervision (by the priest-author, ideally). Affect for sale was not the problem if it met certain standards—namely, that it not promote "sex stimulation" and that it not be wielded within the narrative to create sympathy for crime or immoral conduct (Facey, 90–91).[20] Unedifying spectacles, those that promoted imitation of unacceptable behavior, were not to be permitted.

On one level, then, we can read the interpolated figures of these priest-authors as internalized Daniel Lords, suggesting that films no longer need to be direct transcriptions of religious subjects to do good and to demonstrate the Church's power. But a final powerful irony was in store for the Church. Lord claimed that modernism reduced the Protestant churches to little better than "social agencies" and that the Catholic Church was powerful not only because it alleviated suffering on earth (which even "modernist" Protestants would agree was necessary) but also by the unchanging rightness of its doctrine. Yet the fighting priests noted by the Keysers are popular figures within these fictions not necessarily because their faith in God is unshakable, but because faith helps them run a better social agency. The figure of the priest, in other words, began to circulate in film texts in the guise of a better commodity, just as the Catholic Church was a better social solution for the woes of Hollywood even though Hollywood never endorsed Catholicism's view of "natural law" as much more than a convenience. Clearly this priestly representation is the artifact of the Church's own widening definition of its social mission, a definition that the film industry hardly foisted on it. But if the Church entrusted to the film industry the project of disseminating the message of what Catholicism is for in a multi-denominational society, what inevitably received emphasis was the temporal and the visible—the fighting priest as social agent—with a corresponding lessening of emphasis upon specific doctrinal points or even the necessity of personal faith in God. In short, one could argue that the Catholic Church's association with movies was one of the factors that made it subject to the same commodification to which it claimed the Protestants had given way.

Conclusion

Throughout the period touched on by this chapter, the figure of the Catholic priest or Protestant minister as adviser in the studio was advertised by filmmakers as the hallmark of their concern for religious values, good taste, or the importance of religion itself. The Fox Film Corporation, for instance, reported that in the making of *Thank U* (a story detailing "the conditions besetting the underpaid minister"), "a committee of prominent ministers will be 'on the lot' throughout the time the picture is being produced" ("Ministers to Aid," 429). By the late 1920s, however, an increasingly psychological (as opposed to theological) understanding of the role of religion in social life was beginning to catch up with both clerics and the film establishment. In 1926, writing for the admittedly liberal constituency of the National Board of Review, Andrews presented the connection between film and revival meeting in psychological terms. Here film shares with religion not an interest in sexual titillation but rather the ability to produce a similar rapt or elevated state:

> I could but think in that connection of the way in which the old prayer meetings, revivals, camp meetings and sermons had served just that purpose for the mass of humanity in affording them temporary release from the corroding cares of the day and giving them a prospect of the streets of gold and the gates of pearl.
>
> In a way the motion picture does just this. The beauty and romance of the silver screen help the toilers of farm and factory, of kitchen and office to forget the humdrum of life for a period and come back renewed for the task before them. Any picture which does this is, to my way of thinking, religious. (10)

Andrews's comments are particularly striking because they suggest that religion, like the movies, is a tool of social adjustment. We participate in worship not because we owe a duty to our Creator but because it enables us to go on with our lives. Here is the therapeutic ethos with a vengeance. In presenting religion and film together as necessary forms of psychological manipulation, Andrews's comments look back to the desire to detach oneself from the quotidian found in the juvenile series fiction and Palmer Photoplay Corporation materials. At the same time, Andrews looks forward to the elements of group process psychology (see chapter 5).

If the cleric was a welcome visitor to the set into the 1920s, the psychologist began to take his place both literally and figuratively as the decade wore on. The MPPDA organ *The Motion Picture* reported in 1929 that William Marston, a psychologist formerly of Columbia University, was hired by Universal to "look over story material with a view to injecting proper psychology into them [sic], to study audience reactions, and to do all those things which will better gauge public tastes" ("Psychologist," 7). Perhaps because this article is an artifact of pre-Depression ebullience, it concludes by suggesting that a form of desirable self-fashioning is still available at the box office. Marston is quoted as saying:

> I hope the public will help me by writing me all about themselves, what they get out of motion pictures, and what they would like to get.
> Think this over. Through the Public Service Department of Universal Pictures, YOU, the public, can direct your own emotional development. (7)

The public may not be being encouraged to write films, as it had been previously, but it is certainly being urged to think that in attending them it can "write" itself, and that something intrinsic to members of the audience can still become food for exhibition, if only indirectly.

This article illuminates why the Legion of Decency campaign and the Production Code seemed to resonate with the public for a time. The Depression created a moment in which the film public demanded, more articulately than before, that corporations accommodate themselves to it rather than the reverse. The public nonetheless derived its authority from years of being invited to consider itself in charge of the industry. But the Production Code and the Legion of Decency campaign were only one set of methods for dealing with the corporate menace, as the next chapter will show through an examination of the educational discourses that grappled with the problems of bad film content from a rather different angle. While, as Maltby has argued, the Code substituted for audience research, the impulse toward such research was never far from the surface, and was to return forcefully in the work of George Gallup. It was notably absent from the Catholic approach to film content, however, which rested on moral and aesthetic certainties. But we can see from the comments of Andrews and Marston that a desire to approach the audience more nearly, and on its own terms psychologically, so as to uplift, educate, persuade, or merely sell it things, was the attitude against which the Catholics had to maneuver, and indeed over which they triumphed for a time. Nonetheless, we should understand that the attitudes at stake in the Production Code were to some extent already on the wane by the 1930s, although the temporary consensus they embodied spoke to America's condition in that decade rather more forcefully than did the Protestant establishment, and did so in the lingua franca of all the case studies we are examining—a notion that if film were to be a democratic art form, the public (admittedly variously defined) must be sovereign in its choice of entertainment. The psychological, as opposed to theological, approach to America's entertainment was already gathering strength, as the next chapter will discuss.

CHAPTER FIVE

Learning to Understand the Foe

Character Education and Film Appreciation

What we *like* determines what we *are,* and is the sign of what we are; and to teach taste is inevitably to form character.

—John Ruskin, "Traffic"

As indicated by the discussion in chapter 4 of the complicated relationship of film and organized religion, both Hollywood's foes and its supporters assumed that film held the promise of significant social engineering. Not surprisingly, the industry promoted the idea of a wide range of benefits to be derived from filmmaking and film watching. Under the headline "Increasing General Usefulness of Films," *The Motion Picture* (an organ of the MPPDA) reported in August 1928 that "by-products of the entertainment film" were demonstrating their worth in any number of fields, and that "to possess the motion picture and to deny its use to science, religion, industry, and education is equivalent to possessing the printing press and denying its use to books on science and religion" (Milliken, 6). Any areas of human endeavor that might be said to be concerned with the battle between good and evil were seen as particularly ripe for influence by film. Thus the industry claimed credit for whipping up sentiment against gangsters in the late 1920s and early 1930s; "Deadly Weapon of Ridicule Used by the Screen in Pointing the Moral of Current Crime Pictures Strips Gangdom of Glory," read the subheadline of another *Motion Picture* article ("Debunking"), while Will Hays identified "the American motion picture, produced under the safeguards now in effect in Hollywood . . . [as] a positive and vigorous deterrent to crime" ("Motion Pictures and Art," 5). Film was also sometimes touted as a promoter of significant improvement in international relations. War films, for example, were viewed in some quarters as the best means of inculcating the young with a revulsion against international aggression; thus in a 1930 article in *The Motion Picture Monthly*, Joseph L. Holmes cites two recent studies demonstrating that schoolchildren who had seen graphic depictions of the Great War in film were overwhelmingly disposed toward pacifism.

Although any sustained examination of the uses of film for social amelioration in the context of medicine, industry, or international relations, or even as an educational tool in standard academic subjects, is impossible within the parameters of this chapter, it is important to note that in the 1920s and 1930s, Hollywood actively sought—or at any rate claimed—a role in pursuits that may be considered outside its immediate scope. That the industry was quick to point out the services that film as a medium was performing in a range of areas indicates that Hollywood wanted the widest possible definition of the film medium to shore up its position. Yet again, the boundary between outside and inside was by no means impassable where filmdom was concerned. Indeed, the industry often welcomed being coopted if such cooptation meant increased respectability.

By way of a detailed exploration of what the conception of film as social tool might mean in practice, the present chapter will focus on the film education movement of the 1920s and 1930s, designed for use in high school (and, to a lesser extent) college curricula with two major, often independent aims. The first was film appreciation, in which adolescents were asked to acquire the best possible principles of film selection and interpretation; the second was character education, in which young people were asked to consider moral problems through an examination of short films, and thus come to a better understanding of social and ethical action in their lives. The latter endeavor, arguably, has had the longer pedagogical life, given educators' continuing interest in using film as a way of getting students in middle schools and high schools to discuss moral issues.

But despite the longevity of the practices begun in the 1920s and 1930s, few scholars to date have commented on the beginnings of film education. In this regard, Lea Jacobs stands as a pioneer. In her important "Reformers and Spectators: The Film Education Movement in the

Thirties," Jacobs notes that while character education was the aim of a variety of constituencies, film appreciation was more directly a manifestation of the interests of English teachers, who wished among other things to use motion pictures for intellectual purposes such as the teaching of classics. These two types of film education, however, shared certain assumptions and aims. Both posited that viewer and text shared an extraordinarily close bond, as Jacobs argues; both also hoped to channel and reformulate that bond through a deployment of Deweyan educational principles, as I argue throughout this chapter.

Because they perceived the necessity of fitting text to receiver (a process that resulted from the channeling and reformulation of textual meanings, also discussed in Jacobs), both character education and film appreciation found themselves promoting self-control and self-fashioning through standardized methods, just as juvenile series fiction lauded ostensibly "special" protagonists in formulaic ways and the Palmer Photoplay rhetoric suggested that individuals could stand out from the crowd by following simple rules. These examples illustrate how the Progressive Era struggle against large corporate entities often employed Taylorized or Taylorizing methods. But in no two instances were the stakes identical. In the case of the film appreciation movement, this contest between aims and methods tended to involve questions of individual taste on the one hand and uniform critical standards on the other. In the case of the character education movement, the contest occurred especially between local and national administration of the programs, although here too the individual might come into conflict with standards that character educators presumed general in application.

If juvenile series fiction and Palmer Photoplay offered personal dynamism, educational reformers hoped to generate, through their affiliation with film, social dynamism. The seesaw between standards and something more free-form, and the tension between national and local interests, made the rhetoric of educational reform potentially contradictory about the place of democracy in students' lives. Like the Catholic interventionists discussed in chapter 4, educational reformers appeared to feel that the public had to be organized to express its wishes to the film industry, and it was the task of the teacher, as of the priest, to promote this organization. We may read the methods of the educational reformers as authoritarian to the extent that they foregrounded both a hierarchy of taste and teacher control of increasingly broadly defined areas of student life. Perhaps the most insidious aspect of the use of authority in educational discourse was the deterministic view of human behavior harbored by some educators, who were confident that they understood how film affected audiences and that they could use that knowledge to further social engineering through the schools—an approach that linked the authority figure in the classroom to the scientist and the social scientist outside it.

Both film appreciation and character education assumed that human beings responded to particular film texts in essentially similar ways. It was up to the teacher or guide to bend that response to his or her own educational ends. If Hollywood was mass-producing a one-size-fits-all product that was presumed more attractive than church or the classics, then educators could devise standard (and teachable) solutions to refit the product to the supposedly context-sensitive moral needs of their local constituents.

But one cannot read the educators' rhetoric as uncomplicatedly pro-hierarchy. Some curricula offered to share behavioral knowledge by informing students about the ways motion pictures supposedly created affect, an interest that such education held in common with juvenile series fiction and the Palmer Photoplay Corporation. Moreover, the reformers clearly viewed their interventions as fostering "democracy." The latter perception becomes more intelligible when we consider the impetus these projects received from John Dewey's educational philosophy, which sincerely hoped to bring democracy to the classroom, and which should thus be viewed in the light of larger "good citizenship" programs in the 1920s and 1930s. As Eric Smoodin has briefly noted (23), Dewey's theories both influenced the curricula of film appreciation and character education programs, and inflected the kind of social change that reformers hoped to derive from visual education.[1] Preeminent among his principles were the need to couch learning in terms of activities, the ideal of preparing students for leadership positions, and the desire to integrate elements of the students' extracurricular experience into the curriculum.

If, for film educators, film could be used to reshape the American self, the self could also be used to reshape film. Here the perennial question of affect reappears. The general tendency of the film appreciation and character education curricula was to undermine the autonomy of the film text by insisting on its connection and subservience to other, more authoritative sources of information (such as works of literature or history), or by criticizing film as inadequately concerned with social reality. Observers felt concern over the power of film to provoke emotion in audiences generally, as Jacobs has noted, and thus a corresponding interest in providing young people, in particular, with the means of resisting wholly emotional appeals. Yet, like the religious incursions into Hollywood, elements of the film appreciation and character education enterprises (which occasionally dovetailed) were quite ready to harness affect for their own purposes.

The reformist view that film was a superior means of teaching almost any subject, including ethical behavior, recognized moviegoing's power to change society even as the reformers hoped to reinforce the notion that the film text depended upon the efforts of individual instructors to reach its constituency effectively. The issue was not so much self-commodification on the part of the student (as in juvenile series fiction and the Palmer Photoplay literature) as it was self-commodification on the part of the educational establishment, which hoped to make its traditional product more dynamic and more pertinent to students by affiliating itself with film. At the same time, the film curriculum sought to interrupt the connection of student and film product on two levels: these courses encouraged students to use film as a jumping-off place for other experiences, and also urged them to view participation in film-related activities in school as an essentially democratic activity. The latter view sought to replace the economic individualism of the series fiction and the Palmer materials with a program emphasizing the medium's collaborative nature, a rhetorical trope that reoriented the idea of democracy from the rise of the individual toward responsibility to the group, be it school or nation.[2]

In one sense, classroom use of film was an attempt to assert a measure of local control over the consumption of mass-produced, nationwide products. As will become apparent, however, the local control thus promoted by reformers had its own standardizing and nationalizing tendencies. Both the tension between local and national standards and the concern with and deployment of affect helped to shape a rhetoric concerned with film's place in a democracy. Affect made film a useful social tool, but was intrinsically so powerful that it required a local authority to administer. Concern over affect, in other words, was precisely the opening that educators needed to interpose themselves between a nationwide "image" industry and students' impressionable minds. Yet the educators' methods tended to promote a new kind of standardization in place of that fostered by the film industry.

Reform, Organizations, and the Individual

Whether one views the Progressive movement as having ended with the First World War or as continuing beyond it, the Progressives' contradictory desire for a polity that preserved individualism on the one hand and provided order on the other remained to govern public life in the 1920s and 1930s. Indeed, while the war inevitably heightened the contrast between the two poles, its encouragement of organization capped a trend long in evidence. As Robert Wiebe argues, "national progressivism had been predicated upon the existence of the modern corporation and its myriad relationships with the rest of American society. Chronologically, psychologically, this network had come first. It had set the terms of debate" (296). Peter Conn confirms this view when he states that "the reformers' own objective was to run government *as* a business. The Progressives in particular wanted to adopt the organizational model of the corporation and bring to politics the efficiencies of scale and good management that were enriching the country's bankers and industrialists" (11). Given the corporatist sympathies of the newly professional middle class, their technical training, their perception that they were wedged between the Scylla of uncaring capital and the Charybdis of uneducated labor, it seems inevitable that the solutions that they perceived most readily were those that promised both democracy and order. The formation of voluntary organizations such as the National Board of Review, which mimicked corporate structures but simultaneously offered to mitigate the effects of corporate apathy on the economically and socially disenfranchised, appeared to provide just such a conjunction of order and democracy.

The MPPDA, of course, was tailor-made to reach out to the many voluntary organizations concerned with the place of film in American life, discussed at greater length in the preceding chapter. Ellis Hawley reads the film industry's trade organization as one manifestation (among many) of what he terms Hooverian associationalism, the use of private cooperative ventures as a way of tempering industrial autonomy without necessarily increasing governmental interference (95–96). Citizens in effect assumed some of the burden of monitoring and shaping corporate activity. And, as Morrell Heald has argued, corporations were happy to be perceived as responsive to citizen concern. The rhetoric of corporate service to the commonweal, which of course included not only stockholders and management but also employees and the general public, was to be found in the upper reaches of many of America's most important corporations, such as General Electric (Heald, 130–31). Some of this emphasis on service is traceable to the newly professionalized leadership of America's corporations, in which men trained as lawyers and engineers occupied major management positions (Heald, 131), their educations allowing a common frame of reference with professionals in other fields, such as social workers. While the leadership of the film industry did not, by and large, have professional training (which may be why some Protestant reformers found film leadership at times alien), the MPPDA certainly emphasized the rhetoric of service in publications such as its *Motion Picture Monthly,* which regularly lauded film's contributions to science and art.

This rhetoric of service mandated that the film industry appear to cooperate with all plausible programs of uplift, especially those, such as the film appreciation and character education enterprises, that promised dividends in terms not only of improved public relations but also of increased attendance at certain types of films. Even as the Catholic Church provided the film industry with one solution to its public relations problems in controlling troublesome content at the source (as discussed in chapter 4), the movement to improve content by encouraging better taste continued. When some reformers despaired of the efficacy of direct pressure on the film industry, or found discomfiting the comparatively draconian solution represented by the Production Code— which by no means pleased all critics—they hoped to accomplish their aims by inculcating "self-regulation" in malleable individuals rather than in an apparently unresponsive industry.

John Dewey's Pedagogy as Context

This juncture is where the character education and film appreciation movements enter the picture, and where film, reform, and trends in American pedagogy meet. Insofar as character education and film appreciation took place within the schools, it is necessary to set the stage for the meeting of these various interests. As is inevitable in any discussion of American educational principles of this era, we arrive at John Dewey

(1859–1952). My point in connecting Dewey's ideas to the character education and film appreciation movements is not only to suggest an important context for these programs, but also to make overt the connection between the programs and larger issues in American culture during the 1920s and 1930s, such as education for democracy and the obsession with defining a good citizen. Because Dewey was explicit about the ethical dimensions of education, his reintegration into a discussion of these movements makes more intelligible the concerns and hopes of the movements' ideologues who saw the use of film in ethical terms and did not perceive any gulf between the ethical and the aesthetic.

Dewey was evidently a notoriously difficult writer, to the point of requiring immediate exegesis (see, e.g., Horne). In particular, his work on education apparently resulted in such confusion on the part of would-be disciples that Dewey was obliged to distance himself from various progressive educational shibboleths that others had attributed to him, such as the child-centered curriculum and "contentless" learning. Moreover, a number of the educational ideas for which Dewey is remembered were not original to him, as he would have been the first to admit; he served as the amplifier and transmitter of concepts borrowed from Continental pedagogues, in particular Johann Pestalozzi (1746–1827) and Friedrich Froebel (1782–1852).[3] As a consequence of these caveats, I use the term "Deweyan" with a certain diffidence. In the paragraphs to follow, I mean it to signal those points at which 1930s educational practice appears to be directly deploying an ideal or a principle that Dewey is known to have expressed, regardless of whether he was its originator. By the 1930s, Dewey's refinements to the development of American educational practice were so widespread and influential that pedagogues who were obviously working under his influence (well understood or not) only rarely invoked his name.

Moreover, Dewey was at odds with at least one aspect of educational reform that occasionally impinged on character education and film appreciation. While he was not opposed to positivist approaches to learning, and emphasized science in his curriculum (Ryan, 129), he felt some disdain for reformers who hoped to translate Taylorist models of efficient manufacture from the factory to the school. As Robert Westbrook notes, "The scientific study of education, [Dewey] argued, had too often fueled a deeply conservative impulse to construct a system of education which would simply (if efficiently) reproduce the prevailing social order by preparing students for roles determined largely by their class, race, sex, and ethnicity" (501–2). Hence problem-solving skills were central to Dewey's educational program. For example, children were to learn about wool and cotton "from the standpoint of their adaptation to the uses to which they may be put" (*School,* 18) through handling the actual materials (bolls of cotton and skeins of wool) and considering the practical advantages and limitations of one material over the other, without regard to convention. Dewey reported that his students "worked this out for themselves with the actual material, aided by questions and suggestions from the teacher" (*School,* 19).

As this example indicates, experience has more value in Deweyan pedagogy than has rote learning. The instructor does not present facts for memorization but helps students tease them out of their contact with the environment, and "activity" or "occupation" replaces passively waiting to be filled with information ("if we could take them to the place where the sheep are sheared, so much the better" [*School,* 18]).[4] But Dewey's disregard for conventional vocational education is present in his view that "in the [experimental] school the typical occupations followed are freed from all economic stress. The aim is not the economic value of the products, but the development of social power and insight" (*School,* 16). His curriculum included activities that might appear vocational in impetus because he wished to close the gap between home and school, not because he wanted the school to train children for their future careers. Activities already familiar to children through their extracurricular experience appealed to Dewey, who hoped to use the authority and direction of the teacher to harness children's curiosity about his environment.

Having suggested Dewey's opposition to Taylorism, however, one must also acknowledge the occasional agreement of the two. Concerned with the management of the complex project an experimental school represented, Dewey noted that "the great problem in education on the administrative side is to secure the unity of the whole, in the place of a sequence of more or less unrelated and overlapping parts, and thus to reduce the waste arising from friction, reduplication, and transitions that are not properly bridged" (*School,* 66). The notable word here is "friction," a concept that also exercised Frederick Winslow Taylor, who was desirous of avoiding waste in industrial settings. As Cecelia Tichi shows (75–81), efficiency was dear to the hearts of many figures in liter-ature and philosophy—Henry Adams and William James, for exam-ple—and not alien to Dewey. Dewey's educational program was labor intensive not because it eschewed efficiency but because anything that failed to serve the ultimate goal of real learning was by definition waste: "This question is not one of the waste of money or the waste of things. These matters count; but the primary waste is that of human life, the life of the children while they are at school, and afterward because of inadequate and perverted preparation" (*School,* 59). Taylor could quan-tify efficiency in dollars or man-hours saved, but that kind of efficiency was not to rule over the classroom.

It is easy to see, however, how Dewey's credo of self-realization through contact with one's environment could be perverted into a doctrine of fitting the student to his future place in society. While Dewey's educa-tional program aimed to inculcate leadership skills in any child, regard-less of station or even gender, Dewey was also an advocate of "group-process psychology," a fact that has important ramifications for this study. Because nature and society together constitute the child's envi-ronment, the child will inevitably confront social problems less amenable to utilitarian solutions than are problems in science. Pragmatism, the philosophical matrix of the key Deweyan precept group-process psy-chology (also known as "democratic social engineering" [Lears, *Fables,* 188]), denied the possibility of absolute truths. As William Graebner ob-serves, this attitude "was shared by the social settlement workers, teach-ers, and others who functioned as intermediaries between pragmatic philosophy and the objects of democratic social engineering. Thus group-process theorists and practitioners affirmed the need to avoid the teach-ing of absolutes" (19), an attitude that not so incidentally aroused the disdain and horror of the Catholic Church. In place of absolute truths, moral certainties, or natural law was a "theory of social order . . . guided by a conception of social control in the sense of collective self-regulation and problem-solving" (Joas, 18). Graebner and Hans Joas agree that this notion of "group process" was informed by scientific inquiry and took as its ideal the objective problem-solving techniques and communica-tion of results associated with scientists (Joas, 18; Graebner, 21). Group-process psychology also tended, in the hands of film appreciators and others, to involve researchers in preliminary studies of audience taste with an eye toward using that information to mold audience desires later, an approach that moved this program at times uncomfortably into the domain of market research and advertising in terms of both methodology and aims.

Dewey evidently believed in the primacy of "inquiry" as the "ultimate tool of social action" and as a method "available to, and appropriate to, the smallest and most intimate contexts of life: families, neighborhoods, play groups, the classroom, clubs, and so forth" (Graebner, 21). Group-process psychology rapidly detached itself from its moorings in free inquiry, however, and became adapted to projects far afield from Dewey's and William James's original conception of it. As T. J. Jackson Lears notes, group process was "employed in settling labor disputes, promoting effective salesmanship, and devising schemes for personnel management and progressive education" (*Fables,* 188). This non-Deweyan conception of the function of group process becomes a means of reconciling students, say, to the status quo and obliging the individual to conform to social expectation. Moreover, to the extent that group process results in less social friction, it too promotes a Taylorized conception of social harmony achieved through every member of society finding her or his proper function. In short, group-process psychology potentially returns us to a Taylorist conception of education that is both vocational and concerned with building consensus, possibly at the expense of free inquiry.

Thus Dewey's exegetes could include such Taylorizing spirits, eager to streamline the process of instruction, as Werret Wallace Charters, trained by Dewey at the University of Chicago (Westbrook, 502). Charters is an important figure in the film appreciation movement through his role as superintendent of the Bureau of Educational Research at Ohio State University, where he worked with Edgar Dale (discussed at length below). He was also closely connected to the Payne Fund studies, carried out between 1929 and 1933, in which a scientific approach to social problems and film becomes most evident.[5] Both the master/pupil relationship and the philosophical divide between Dewey and Charters are significant because, together, they suggest how the film appreciation curriculum could be Dewey-inflected without necessarily being something that Dewey would have claimed as his own. Indeed, Charters's influence on the film appreciation curriculum did not prevent Dewey's democratic ideals from informing the movement, inasmuch as Dale hoped that films would become more responsive to actual social conditions and thus expose social inequalities, an aim that Dewey would have ratified.

Film Appreciation

It is hard to determine precisely how many people were involved in organized film appreciation, but it is safe to say that it was a fairly widespread phenomenon in the early to mid-1930s. The academic literature on the subject is substantial, and film appreciation effectively had its own periodical (*The Motion Picture and the Family*, produced with the aid of the MPPDA). The movement's own figures suggest a rosy picture of activity and progress. For example, *The Motion Picture and the Family* reported in October 1935 that 2,500 schools had adopted William Lewin's appreciation curriculum ("2500 Schools," 1). Cline M. Koon reported in 1934 that the National Council of Teachers of English (NCTE) had sent study guides for selected motion pictures to 17,000 English teachers. Koon also noted that "1,851 pupils in 19 states, 31 schools, and 57 theaters in 28 different cities participated" in 1934, one of the movement's first years (6).

In practice, film appreciation could be an autonomous academic subject, a project undertaken by a community group or a club affiliated with a high school (often serving as a reward for academic excellence), or a unit within an English class curriculum. *The Motion Picture and the Family* reported that "before a boy or girl can gain admission to the Cinematography Study Group [at East Orange High School], which has a membership limited to 27, he must have attained a general average of B in all his studies, plus an A average in English" ("Motion Picture Study," 1).

In addition to *The Motion Picture and the Family*, a wealth of advice and instruction addressed those interested in establishing a film appreciation club or sequence within a high school or extramurally. The two major producers of publications were Educational and Recreational Guides, Incorporated, published out of Newark, New Jersey, and the Bureau of Educational Research at Ohio State University. Lewin was the presiding spirit behind much produced by Educational and Recreational Guides, which appears to have addressed itself partly to the significant retail possibilities latent in the prospect of a national film appreciation curriculum, inasmuch as the major focus of Lewin's publications was the student or teacher who might subscribe to a series of guides on specific films.[6] Lewin was affiliated with the National Council of Teachers of English and wrote his doctoral dissertation on the subject of measuring improvements in student taste attendant upon completion of a film appreciation curriculum. Much of the material emerging from Ohio State, on the other hand, was addressed to the theory of film appreciation and studies of its efficacy or general instructions on how one might go about establishing programs for children or adults in the community. Dale, the most notable figure associated with this aspect of the movement, wrote numerous scholarly articles on the subject and prepared a general textbook (published by Macmillan as a volume in the Payne Fund studies) on the subject of film appreciation, *How to Appreciate Motion Pictures*.[7]

The guiding spirits of the film appreciation movement in American high schools saw film appreciation as three things: an activity, a skill, and a form of consumer behavior. Film appreciation hoped to correct the increasingly passive individual behavior that reformers saw as governing American leisure and labor. Moreover, Dewey's educational philosophy emphasized the centrality of "experience" both in education generally and in aesthetic education particularly. To the extent that film appreciation could present itself as an activity, it offered students the "experience" some critics felt was lacking from film attendance per se. In Deweyan terms, teachers could bridge the gulf between home and school through a curriculum that leveraged the child's extracurricular experience into learning and reformulated classroom work into hands-on activities rather than mere soaking up of knowledge.

The film appreciation movement was above all interested in providing a set of standards by which to judge films in both moral and aesthetic terms. As a branch of learning, it closely resembled similar curricula in music or art appreciation in its emphasis on developing taste and capacity for value judgments. Some critics went so far as to assume that film was a branch of literary activity and thus belonged within the English curriculum. Walter Barnes, for one, firmly pronounced that "we can appraise the photoplay accurately only when we discuss it as literary art" (2). Film was often acknowledged to be a significantly more popular art form than literature, but one that had not yet produced a great number of classics; its prevalence in students' lives suggested its place in the modern curriculum. Deweyan principles appeared to urge the inclusion of new studies to replace subjects that no longer reflected the child's experience outside the classroom.

Including film in an academic curriculum participates in Dewey's belief that "[there is no gap] in kind (as distinct from degree) between the child's experience and the various forms of subject-matter that make up the course of study" (Westbrook, 99). Clearly, the project of connecting education to children's lives mandated the recognition of leisure activities. To incorporate these activities into the classroom was for Dewey a way of acknowledging the importance of all the child's pursuits, but it also brought them under the purview and control of instructors, who could then subordinate everything, including activities formerly defined as play, to education.[8] Dewey held that "the question of education is the question of taking hold of [the child's] activities, of giving them direction. Through direction, through organized use, they tend toward valuable results, instead of scattering or being left to merely impulsive expression" (*School*, 37–38). Promoters of film appreciation thus designed their curriculum to restore rigor to an activity that had long been, at least in the minds of its detractors, devoid of skill. Under the control of qualified instructors, commercial leisure could be the means of drawing out important capacities from children.

As a consequence of the application of Deweyan principles, then, the doyens of film appreciation (preeminent among them Dale and Lewin) often focused on the discrete bodies of knowledge required to form a serious opinion of a film's worth. Like correspondence schools, film appreciation texts served up the recondite vocabulary that film appeared to require as a new technological and industrial art form, defining for readers terms and concepts such as "angle shot," "inkies," and "lap dissolve" (Dale, *How to Appreciate*, 235, 237). But if correspondence schools offered such information in the interests of promoting students' ability to become film industry insiders, film appreciators provided such knowledge under rather different terms, seeking to achieve seamlessness between academic preparation and the world of vocations that students would grow up to explore.

Film appreciation was not exactly vocational training, however. Instead, it reflected Dewey's concern with "occupation," which was "a mode of activity on the part of the child which reproduces, or runs parallel to, some of [the] work carried on in social life" (quoted in Westbrook, 101). Film appreciation thus mimicked real-life behaviors such as film reviewing and, to a lesser extent, film production, but was not necessarily designed to prepare students for careers in these fields so much as to suggest what was wrong with an uncritical endorsement of Hollywood products. While the Catholic approach (discussed in chapter 4) desired to separate the viewer entirely from evil influences, educational reformers sought rather to arm their constituents against unhealthy discourses. Lewin, probably the most commercially minded of the film appreciators, was also the readiest to point out the value of photoplay clubs and allied activities in helping children "choos[e] vocations through these experiences" (Jackson, 8), a viewpoint that presents vocational training as a secondary benefit of affiliation with the film appreciation curriculum. Indeed, Lewin's concern with vocational training crops up again in a somewhat different context with his use of industrial films designed to "emphasize the life-career motive" in a composition course apparently about occupational choices. Lewin lauds this use of film for its greater educational efficiency (Lewin, "Photoplays," 454).

While few film appreciation teachers were unrelentingly hostile to the industry, they did not invariably share the basically sanguine outlook of their rivals operating out of Hollywood, such as Palmer Photoplay.[9] One area where Palmer's methods and the film appreciation movement's preoccupations meet involves their attempt to educate an audience presumed to be uninformed. In the case of the film appreciation movement, this approach follows from Dewey's concern with the worker's increasing loss of skills. Without necessarily inviting pupils to consider themselves future members of the industry, film appreciation sought, in some measure, to restore to the worker (the audience, in this instance) the skills that management—typically the producers—might otherwise hoard or disguise.

The emphasis upon the craft necessary to make a motion picture not only suggested the skills required; it also had the effect of replacing one kind of knowledge about motion pictures (fan knowledge) with another that may have struck reformers as less uncritical. Typically, in a form of proto-auteurism, film appreciation writers tended to locate the control of craft with the director. Dale makes the point about hierarchies of knowledge explicit in *How to Appreciate Motion Pictures:*

As a matter of fact, a knowledge of a star's director is more important to you from the standpoint of an understanding of motion pictures than is a knowledge of what the star wears in the afternoon, whether she likes buttered toast, the number of hounds she has around her place, and the other things that motion-picture magazines publish about stars.... making motion pictures is a cooperative enterprise in which stars, directors, other members of the cast, and technicians must share honors. (179–80)

Indeed, awareness of the role of the director was one criterion for progress in a film appreciation course. F. P. Frutchey and Dale note that "changes in the behavior of high-school pupils who had the course in motion-picture appreciation were shown by the fact that before the course 28 per cent of the pupils indicated that directing was of much importance, and at the end of the course 68 per cent indicated that directing was of much importance" (35–36).

This industrial orientation becomes clearer when we realize the extent to which the film appreciation movement resembled other Progressive Era projects designed to protect consumers. Film appreciation's ultimate aim was to develop a kind of consumers' union among the rising generation of film attendees to force certain concessions in content from management. It would effect this change through the promotion of activities that, while connected with film, to some extent competed with and supplemented the act of going to movies. For instance, to limit the power of illusion and also to suggest that filmmaking was a body of knowledge with craftlike aspects, which the novice appreciator must learn, it accentuated the craft and skill behind film production and foregrounded the relationship between the knowledge of these arts and the information required to be a "good" consumer of the finished product, whether or not such consumers intended to have any contact with the film industry beyond purchasing its products.

It was through contact with young people (and adults) as consumers-in-training that the film appreciation movement had the greatest potential economic effect on the film industry. The film appreciation and character education movements did not foster censorship but rather hoped that their programs would result in such complete internalization of the principles of film selection that the market for inappropriate films would wither on the vine. If the film industry practiced "self-censorship," so would the ideal pupil. Here the film selection movement urged, consistently, by more liberal Protestant or ecumenical groups such as the National Board of Review triumphed over film censorship as such. Thus two interlocking programs of "self-censorship" coexisted, one affecting the industry and the other affecting the good citizen, and that reformers had been pursuing both solutions, although not every group endorsed both methods. For example, the film appreciation movement was generally liberal in its sympathies. A "Ten-Point Film Program," a kind of manifesto for the film appreciation movement, appearing in a *Group Discussion Guide* for the 1938 picture *Men with Wings,* vowed "to oppose legislation which attempts to solve the cinema problem by artificial censorship, as against that 'natural censorship' which comes through education" (Doyle, 13), suggesting that it was by no means only Catholics who opposed national censorship legislation.

Self-censorship of either variety also obviously marks the point at which reformist concern has shifted entirely to texts and away from venues as the source of any potential moral problem. Film appreciation was never an attack on exhibition, in other words. Instead, its object was the elevation of viewers' taste, a goal by no means inimical to the interests of independent exhibitors. If teachers could deliver on their promise to pack movie theaters showing good films, exhibitors would reap both goodwill and profits. While the classroom was the zone in which students were to be educated about films, school was not actually a rival exhibition site.[10] Even the goal of standardizing experience could be promoted through attendance at the local theater, as Lewin notes in the *Teacher's Key to Accompany the Student's Guide to Critical Appreciation of the Photoplay Version of Stevenson's* Treasure Island: "An ideal plan for seeing the photoplay is to let the class go to the theater in a body, with a view to discussing the picture the following day. Unless all members of a class have seen the picture, and preferably at the same time and place, participation in the discussion is hampered by variable factors beyond the teacher's control" (3).

Exhibitors frequently put their theaters at the disposal of groups of educators interested in film appreciation. For example, according to Donald Eldridge, a New Haven film appreciation group profited from the availability of reduced-rate admissions, secured whenever a teacher accompanied a group of students to their local theater (176). Clearly, the local theater was the main exhibition venue for the screening of the basic material of the film appreciation class. Teachers returned the favor of cheap admission by providing a semi-captive audience of students ("semi-captive" since film attendance was never mandatory). Moreover, the issue of block booking was often taken up in the classroom (Eldridge, 177; Pollard, *Teaching,* 5),[11] which meant laying before students those aspects of the industrial structure of American filmmaking that appeared to constrain the ability of local theaters to play the films that community leaders wanted featured. Indeed, film appreciation materials were essential to the marketing of big-budget adaptations that might be comparatively difficult to sell to some audiences; as Richard Maltby notes, study guides "could be used by distributors to justify the film's exhibition in neighborhood and rural areas where such films were frequently canceled" ("Code," 63).

But neither the movie theater nor the individual film stood alone in the ideal film appreciation curriculum. Film appreciation worked by insinuating itself into the connection between Hollywood and viewer as an indispensable middleman. Significantly, Deweyan pedagogy often had the effect of diminishing the autonomy and authority of the film through strategies such as presenting classroom discussion as essential to understanding; providing production information and, in the case of period films and adaptations, historical or literary background; and emphasizing activities designed to lead students away from the consumption of the film as an autonomous work of art toward a contemplation of its content, of which film may not be the only or even the best purveyor. Indeed, the credo of the film appreciation movement appeared to be "Start with the film in order to get your students to read or think."

Lewin's career is instructive in this regard, and suggests the ways the film appreciation movement acknowledged the dependence of older subjects on the glamour of film to make instruction exciting. An account for young people of the inception of his film appreciation program at Newark's Central High School emphasizes that the Photoplay Club was the ideal venue for engaging with texts that might otherwise have little savor. *Ben-Hur* appears to have been the work that energized this process, as, when one of Lewin's pupils hoped to substitute a movie report for a book report (Jackson, 8), Lewin insisted that the student read the book as well in order to be able to assess "whether the picture has told the story faithfully and preserved its beauty—and if it has, how it was done." Thus as early as 1927, the date of this anecdote, Lewin took the position that consumers of films should judge them with reference to authorities outside the film text—to the literary or historical models that the best films would inevitably follow. Such invocation of outside authority inevitably placed the film in an inferior position as it tried, probably unsuccessfully, to measure up to its source; even the most popular print texts became part of high culture in comparison. Such an approach rendered film safer by forcing consumers to acknowledge its cultural subordination, even while the charisma of the medium was supposed to persuade those same consumers to read books.

Generally speaking, film appreciation activities ranged from simple membership in a group such as Newark's Photoplay Club, to participation in pursuits in which the experience of film attendance morphed into other forms of knowledge. For student members of film appreciation groups, potential activities included not only writing reviews but also "the use of a bulletin board,[12] a trip through a projection room, a visit to the stage back of the screen, inquiring of the exhibitor as to future bookings, and writing letters to the theater manager" (Lewin, *Appreciation,* 70). Students who saw *Mutiny on the Bounty,* for instance, might have had to write "a newspaper article setting forth the approaching voyage of the *Bounty* for the obtaining of breadfruit in the South Seas and the transporting of it to the West Indies. In the article tell something about the personnel of the vessel" (Law, 28). Students seeing *Captains Courageous* were to "conduct a ship model contest" or "make a map showing the coast of New England, Nova Scotia and Newfoundland, and chart the probable course of the *We're Here*" (Orndorff, 13). And students working on *Victoria the Great* could make "costumes for Victorian characters who appear in dramatization. (Materials for paper costumes may be secured from the Dennison Manufacturing Co., Framingham, Mass.)" (Fowler, 12).[13]

Such activities, of course, did not typically take place on a highly emotional plane, and, as Jacobs notes, many members of the film appreciation movement hoped to frustrate the process of identification presumably operating in the enjoyment of film texts (32). Jacobs attributes the desire to minimize the emotional control that the text wielded to mistrust of the too-emulative audience, likely to be swept off its feet by displays of spectacle and high life available only in theaters. She supports this interpretation of the major tenets of reformers' notions of spectatorship by recourse both to the Payne Fund studies and to those aspects of the film appreciation curriculum designed to thwart uncritical acceptance of the filmic text by holding the photoplay to high standards of truth-telling and autonomy (38).

In fact, however, what one finds from another survey of materials surrounding the film appreciation curriculum is not a rejection of spectacle and emotional arousal per se so much as an interest in well-defined kinds of spectacle in the service of historical pictures and literary adaptations. Jacobs's valuable observations on the centrality of "plausibility" and the minimizing of spectacle notwithstanding, perusal of *The Motion Picture and the Family* suggests a substantial interest in the lavish film derived from literary or historical sources that inevitably involves a kind of spectacle.[14] Two factors may explain this unexpected emphasis: first, an avoidance of manifestations of everyday high living, with these narratives safely located in the past, and, second, a desire to harness the storytelling power and authority of the motion picture to more purely pedagogic ends. Thus we find not only the emphasis on plausibility that Jacobs identifies, but also a corresponding highlighting of the necessity of good research for motion pictures set in the past, and of the faithful screen adaptation of works whose source is literary—both of which approaches steer the attention of the student away from the film text as such. Jacobs attributes to the film education movement a pronounced interest in undermining the authority of the "unrealistic" film text in favor of more "plausible" or "truthful" film texts, which is undeniably the case. But the film appreciation movement goes further in its systematic questioning of the autonomy of even the best film texts by constantly referring the viewer to their sources, whether in fiction or in reality.

In discussing plausibility, it is important to bear in mind that film texts were held, in some instances, to remarkably high standards of social reality. For example, Helen Miller Rand asks, "What is the proportion of rich men and poor men in this country? How does the proportion of rich men and poor men in moving pictures give a fair picture of economic conditions as they actually exist?" ("Motion Pictures," 165). Moreover, Rand suggests (in an attitude that Dale shares) that the present sociological understanding of crime must be manifested on the screen: "*The Devil Is a Sissy* showed that a bright boy is made into a criminal by his environment. That is constructive teaching. When we realize that slums make criminals, we know that the real criminals are those who make slums" (165).[15] Barnes offers a more sensitive critique of the plausibility question when he argues that

> it can be said at once that in historical authenticity, in costuming, furniture and furnishings, in place settings and local color, in manners, customs, and language, in all the details that make for fidelity to the objective facts, the modern photoplay is admirable.... But when we turn our attention to another kind of truth-telling, faithfulness to the facts of human nature, we perceive immediately that the photoplay has not been so successful. (30–31)

For Barnes, the psychological unreality of such figures as the fallen woman "more sinned against than sinning" undoes the visual reality of well-prepared art direction, although he attributes this failing in part to producer responsiveness to audience demand for "sentimentality" (32). Given Barnes's distrust of American taste, a view (as argued in chapter 4) by no means unique to him, a program that addresses the taste problem at its source seems exactly the one needed.

If anything, film appreciation materials suggested that industry self-censorship was responsible for the avoidance of more probing and socially responsible stories. According to this view, the Production Code offered producers an easy way to avoid discussing hard issues. The *Group Discussion Guide* for *The Citadel* provides a stinging comparison of Hays Office standards for acceptable motion pictures to those of the Institute for Propaganda Analysis. For the Production Code's "Does [the film] observe the conventional taboos as to indecencies in matters of sex?" is opposed to "Does [the film] avoid the implication that the successful culmination of a romance will solve most of the dilemmas of the hero and heroine?" (Dugan, 13). Similarly, *Guide*'s author clearly considers the MPPDA instruction to "avoid expressions of profanity and scenes of excessive drinking" inferior to the Institute's prescription to "avoid the implication that the good life is the acquisitive life, with its emphasis on luxury, fine homes and automobiles, evening dress, swank and suavity" (13).[16]

Dale joins Rand, Barnes, and Lewin in demanding more social reality from motion pictures. "I want a dramatist to show the ruthlessness and cruelty that lies back of much of our so-called urbanity, our well-fed amiability," he writes. "We need film-makers to show us why America lynches Negroes, why hate comes so easily, why love is so cheap and sentimental" ("Teaching," 119). The subjects that Dale demands would flout the Production Code, violate Hollywood's lore about what would and would not play in certain regions of the country, and probably make even sympathetic Catholics cringe because such stories would tend to employ a naturalist idiom inevitably showing characters defying "natural law."[17]

Dale's views here suggest that the connection between film appreciation and character education is that in both, teachers cause narrative film to serve as a guide to conduct. In considering the purpose of film as an art form, Dale argues in *How to Appreciate Motion Pictures:*

> Through the motion picture we can have displayed for us the finest responses that our finest individuals have been able to make in certain important decisions of life. In other words, it can provide patterns for the highest and most intelligent conduct of which man is capable. It can show us characters trying to discover what is most valuable in life and the methods which they use to achieve their goals. (207)

More pithily, Dale states elsewhere that "the question 'What standards shall be used in evaluating motion pictures?' is really the question 'What is the good life?'" ("Teaching," 115). But while the photoplay may be a laboratory for the exploration of choice, some film appreciators expect responses to films to form a remarkably regular pattern.

Given Lewin's belief that "a preliminary study of the problems involved in any plan for teaching boys and girls standards for selecting and judging photoplays reveals at once that the photoplay is the most complex and the least standardized of all mediums of expression" ("Standards," 799), it is perhaps unexpected that film appreciation required students to express their feelings about films by standardized rating scales. Such scales gathered information about both films and pupils and thus actually constituted a form of audience research designed to forward the project of consensus on the value of individual films.[18] The desire to collect information about audience responses to film links this reformist discourse to those discussed in chapters 2 and 3. The Catholic approach that dominated the Production Code was wholly uninterested in gauging audience response. But any discourse that made popularity the means by which the audience was to be reached for some other purpose automatically enshrined even a debased group-process approach.

Ostensibly, regular completion of the rating scales would afford a record of progress in the development of taste and standards for judging motion pictures. In fact, of course, the scales would inevitably have the effect of enforcing or at least urging a certain uniformity in standards, regardless of picture and regardless of student. Rating scales went through changes over time, and were tailored to particular grade levels. Typically, the forms asked the pupil's name, age, grade, sex, and school; with reference to films, they asked the title, producer, author, director, and star. The student gave numerical scores for such elements as main idea ("Important v. Unimportant"), story ("Illogical v. Logical"), characters ("Artificial v. Lifelike"), photography ("Ordinary or Crude" v. "Unusual or Beautiful"), and value to humanity ("Destructive v. Constructive") (Lewin, *Photoplay Appreciation,* 90). More elaborate scales for higher grades of students included "Voice of Star," where the possibilities included "Annoying, Defective," "Rather Uncultured," "Not Very Noticeable," "Effective," and "Remarkably Versatile." For dialogue, students could choose among "Trite," "Colorless," "Rather Witty," "Clever," and "Brilliant" (*Photoplay Appreciation,* 92).

What is intriguing about these rating scales is that they simultaneously address moral issues, such as the scope of the theme and the picture's value to humanity, and more trivial elements, such as how cultivated the voice of the star happens to be. The scales emphasize logic and plausibility, yet prize most highly the themes described as "Momentous, Epical" (*Photoplay Appreciation,* 92), suggesting again that spectacle has a place in the film appreciation curriculum and in the hearts of students. We must bear in mind, of course, that scales deploy language that cannot fit every picture. Yet the rating scale also crystallizes the standards that the sponsors of the movement want the student to apply. The scales oscillate between rejection and endorsement of Hollywood's more seductive wares. Clever dialogue is associated with pictures set in Manhattan apartments, not with those set among sharecroppers; the emphasis on beauty in photography and composition also potentially moves away from the depiction of reality. While it is possible that this confusion of aims results from poor word choice on the part of the authors of the scales, or from the exigencies of compromise within committee work, the confusion may also arise from the contest between two opposing attitudes toward art. On the one hand, works of art expose us to the beautiful, which should uplift us morally; on the other, they reflect reality, not always so uplifting.

Here film appreciators are caught between the twin objectives of their curriculum, to expose students to experiences that are intrinsically "beautiful" and to enlarge student sympathies through education for citizenship. As Lewin observes in his foreword to Sarah Mullen's *How to Judge a Motion Picture,*

> the moving-picture appreciation movement combines very well with two new educational trends—the teaching of the proper use of leisure time and the new emphasis on social attitudes in the teaching of literature. If our millions of high-school students learn good standards for judging the photoplays which they will inevitably see, there is likely to be at least a slight advance along the entire front of human thought. (Mullen, 6)

Film appreciation is thus part of a larger project of social hygiene within American schools, on a par with "character education, health education, vocational education, art education," according to Mark May (159).

Clearly some instructors felt diffidence in manipulating rating scales. Leon Mones, in his guide to *Music for Madame,* describes two schools of criticism, the sum-of-the-individual-parts school and the "whole impression" school.[19] The first school applies its standards individually: "These critics . . . determine the value of a photoplay by making an inventory of its theme, plot, educational value, skill in direction, acting, photography and lighting effects, setting, costumes, properties, sound and music, dialogue, social and moral worth, etc." (11). While Mones introduces the possibility that the whole-impression school may have something to offer, or that a combination of the two approaches may be a good strategy, he nonetheless provides another scorecard for his readers, saying, "we should also be able when we desire to analyze and tabulate with some degree of logical and scientific accuracy the elements of an artistic experience" (11).

It would be all too easy to parody these educators as latter-day Gradgrinds wearing down a generation of Sissy Jupes, but at least they paid popular culture the compliment of taking it seriously. They were also, of course, trying to bend it to their own pedagogical ends in an effort to move with the times. It is possible that the scorecard appeared at just the moment when educators threw up their hands at the task of teaching taste in the multicultural classroom of the 1930s. Also in keeping with the Gradgrind mentality, a certain classism, both overt and covert, was at work within film appreciation, as Jacobs has discussed in detail in "Reformers and Spectators." One of the less rigorous texts associated with the project asks students, "What is the effect of too much make-up in a picture? In life?" (Sterner and Bowden, 30). Of course, the entire enterprise of uplift could not be too overt or it would risk exposing class biases. Elizabeth Watson Pollard cautions teachers that "care must be exercised to prevent any feeling that highbrow standards are likely to be foisted upon the group. Stress greater enjoyment rather than greater appreciation" (*Teaching*, 23–24). So confident in their own judgment and the efficacy of their teachings were reformers that they assumed, as Pollard also notes, that "the cheap, the tawdry, the sensational, if presented in the true light, need not be directly belittled.... When poor films are presented without prejudice, the consensus of the class will favor the better things" (*Teaching*, 30). In a way, then, we are back to the optimism of the series books and correspondence school manuals, the conviction that public taste is reliable. What differentiates the present discourse from these two predecessors is that it is talking about an *educated* public taste.

Of course, taste is not a matter of mere individual preferences for film appreciators, nor is it a simple class marker. Rather, it is a weapon in the ongoing box-office plebiscite: "Students should be made to realize that they are voting for the continuation of low-grade pictures every time they attend a poor picture" (Pollard, *Teaching*, 38). Consumption and production participate in a feedback loop so sensitive that if teachers or other authorities act only on the audience, they can improve the quality and content of films. In a refinement of the linkage between consumption and production also seen operating within children's series fiction and Palmer Photoplay, here is the perfect identity between consumer and product. When Pollard, in her guise as promoter of yet another rating system, asks her adult audience to consider whether a film contains "any elements inserted for the purpose of attracting a certain type of box-office patron," she is highlighting the problem of Hollywood's pandering to sensation seekers (*Groups*, 50), whom film appreciation must retool into more refined consumers.

Clearly the project of social engineering or social hygiene will be most successful with the young and malleable. While some film appreciation texts assume adult interest, or the necessity of uplifting adult taste as well as children's taste, children were nonetheless the main targets, as in other social hygiene enterprises such as "Americanization." Further, in an interesting reprise of the lessons of the series fiction, in which the presence of children in the film industry demonstrated its domestic nature, film appreciators gloried in being able to develop in children standards so rigorous that adults trembled before the judgments of the young Solomons thus created. Mary Allen Abbott reports with pleasure that "recent studies show that the Horace Mann children have an even stricter standard than the [adult motion picture] committee as concerns good taste and lack of vulgarity!" (quoted in "Raising the Standards," 12). Phyllis Jackson delights in noting "that a film has been acclaimed by the critics or has won a national award does not mean it will pass the judgment of the [Newark] Central High Photoplay Club" (10).

The metaphor that film appreciators deployed to explain how they hoped to undermine the attractiveness of poor films was "inoculation," which continues rhetorically the strategy of reformulating the audience at the earliest possible age. Frutchey and Dale present "immunity to extreme emotional reactions and undesirable attitudes commonly produced by motion pictures" as one of their thirteen objectives of motion-picture appreciation (34). Melodrama in particular was the code word for the "undesirable," presumably because it partook simultaneously of low-art subjects and unrealistic presentation. Writers often used melodrama and the unfortunate pleonasm "overexaggeration" interchangeably to describe what the film appreciation curriculum particularly wanted to eradicate in both student taste and film texts. Lewin expresses his hope that "we shall emphasize the intimate, the subtle, the charming, rather than the melodramatic" ("Standards," 808), while Abbott makes the point that the standards being developed apply to adults and children alike, since the standards deal with such general issues as "suitability of subject matter, with good taste in comedy, and with lack of over-exaggeration of emotion" ("Classics," 627). Dale, ever the most articulate and sensitive of the ideologues behind the movement, makes explicit the ultimate aim of the denigration of melodrama: "We are trying to extend the range of living that is governed by rationality and perhaps decrease the range that is governed by emotionality" ("Teaching," 114). The inoculation metaphor operated not only with regard to the rejection of melodrama but also in regard to what Dale evidently called "sensitizing pupils to important social problems" (Smith and Lemon, 207), a project visible in some attempts to teach social studies through film production.

The logical extension of film appreciation would be into film manufacture, which certainly occurred in schools, although it is hard to say how frequently or with what success. Many film appreciation curricula (possibly under the influence of notions of vocational education) assumed that if the resources could be mustered, boys in particular would be eager to learn to make films. The whole discourse during this period surrounding amateur filmmaking is a subject too large for this chapter, but we note the presence of one fascinating example important to our understanding of the *practice,* as opposed to the *theory,* of film appreciation.[20] At the Horace Mann School in New York City, one of the major sites of film appreciation on the East Coast, the girls' tenth grade engaged in film production in their social studies course. In 1937, nine girls (with faculty and technical help) made a film called *Their Adopted Country* about the experiences of an Italian-American family. Given the emphasis on the avoidance of cliché in writings about film appreciation, one is a little startled to discover that the characters include a talented daughter who is befriended by wealthy New Yorkers on her way over from the old country (she entertains them with her violin, which is later stolen), a brother who runs afoul of the law when he becomes involved with bootlegging, and a mother who feels rejected when her daughter prefers American fashions to Italian fashions (Smith and Lemon, 209–10).

One of the stated results of the project was a greater appreciation of Hollywood's efforts (Smith and Lemon, 212), in keeping with Patricia Zimmermann's observation that amateur-film discourse "educated its users toward a better understanding of commercial filmmaking processes" (51). Zimmermann argues that such rhetoric had two traits: first, the promotion of amateur participation in "conventional, commercial cinematic practices" and second, the representation of such mastery as the means to social rise via the transformation from amateur (a somewhat scorned category) to professional (the standard by which to judge all film). While, from a perusal of *Amateur Movie Makers,* adult amateur productions appear to have exhibited the rhetoric of social rise on which Zimmermann comments,[21] this discourse tended to be less apparent in discussions of such activities in schools, where the purpose of such projects in humanities classrooms was almost never vocational. Instead, we find that school filmmaking was the occasion for visits to historical sites, intensive research, creative writing assignments, and the exercise of leadership by a large range of students. Filmmaking projects, which cropped up outside the film appreciation curriculum as a means of mastering an unrelated subject, such as ancient history, directed attention away from the more problematic aspects of Hollywood, such as the manipulation of stars, and toward previously neglected categories of labor, such as editing and direction. Ultimately, of course, one function of student filmmaking was to inoculate students against a passive acceptance of filmic illusion by urging them to explore how films are made.

Student filmmaking, in short, represented the Deweyan "occupation" par excellence. Not only did it offer direct contact with problems in the representation of history and the construction of narrative, but, as a collaborative activity, filmmaking also served as an ideal field for the operation of "group-process" psychology. If *Their Adopted Country* was a thematic exploration of some of the problems of democracy in the immigrant experience, its sponsors also viewed the process of making the film itself as an exercise in democracy: "The entire project was handled on a democratic basis. When the idea of producing a movie was first suggested, the group was immediately enthusiastic; but many valid arguments pro and con were brought forward in class discussion, before the final affirmative vote was cast" (Smith and Lemon, 207). Dale summarized an account of filmmaking at New Jersey's Montclair High School in which the author suggested that student films were the ideal successor to the old autocratic ways of teaching drama: "Formerly the teacher chose the play (usually some 'classic') and picked out the 'best' actors, who then put on the play. The newer methods, however, bring in a variety of student activity, and include the production of short plays and one-act plays, even of student-written plays and films" ("Production," 272–73). The resolution of both technical and social problems resulted in self-discipline and resourcefulness, a theme that often cropped up in Dale's summaries of articles about student filmmaking in "Film Production in Schools." The school film production exercise at least implicitly hoped to foster an extratextual narrative of filmmaking as collaborative endeavor and democratic enterprise (albeit one conducted under the authority of a teacher) to undermine the construction of stars as individuals elevated at the expense of the mass.

The appreciation of Hollywood's products that one may gain from these experiences is not necessarily an endorsement of conventional narratives. While one result is obviously an increased respect for the technical and managerial difficulties inherent in filmmaking, another is clearly the sense that the skills involved are masterable, leading to a "keener and more critical attitude toward the motion pictures they [students and faculty] see in theatres" (Dale, "Production," 279). Here there is an erasure of boundaries between filmgoer and filmmaker of a kind already on display in chapters 2 and 3, albeit without the vocational cast promoted by the juvenile series fiction and Palmer Photoplay Corporation. Yet although the film appreciation movement marshaled hostility against the deployment of cliché and coincidence within Hollywood films, even *Their Adopted Country* fell short of the ideals of the movement, suggesting the power of the Hollywood model in practice. What one discerns here is the inescapability of the manipulation of affect as a tool for the discussion of social issues, even allowing for the possibility that teachers provided such knowledge unintentionally: "If this original scenario is alive with social implications, the student-dramatist will probably be more alert in examining the movies as a social instrument and in detecting significant themes in current productions" (Smith and Lemon, 212). We find ourselves confronting within the film appreciation movement, at least within this example, precisely the same gap between precept and example that we found operating in the case of the Palmer Photoplay Corporation, which also denigrated melodrama even as it appeared in the products that bore the Palmer imprimatur.

Melodrama, of course, was not just a matter of the mode of the narratives that students watched or made. It was also the stuff of the discourse surrounding the presence of film in their lives. Film, as the Production Code acknowledged, is a potent medium, and appreciators described it at times as if it were a terrible force of nature that humankind must learn to control. Barnes, whose judgments were generally quite reasoned, himself gave way to melodrama in describing the effect of film upon the senses of the viewer:

> Applying these general statements concerning restraint of emotions in fine art to the average, or even the superior photoplay, we are impressed immediately by its excessive display of emotions, its frontal, often brutal

attack upon our senses, by its failure to temper and discipline itself as an
art-form. Perhaps this failure is due to deliberate design on the part of
the producers: they feel that their audiences desire and demand nothing
more, every night in the year, than saturnalias, orgies of sensational
indulgence. . . . The careful, sensitive artist . . . recognizes the intrinsically somatic nature of his medium and tones down his effects
proportionately. (21)

If film assaults the viewer somatically, as Barnes puts it, it also assaults
the viewer morally, as Henry James Forman would have it throughout
his summary of the Payne Fund studies, *Our Movie-Made Children* (1933).

Conversely, the terms critics use to praise the potential of the moving
picture are themselves melodramatic. Forman exclaims that

> The millennial dreams of all the saints and sages could scarcely have
> aspired so high. Here is an instrument fashioned at last in universal
> terms. Send forth a great message, broadcast a vision of truth and
> beauty, if only you broadcast it by means of the so-called silver screen,
> [and] literally all America will be your audience. Who could have
> imagined a population more nearly, more inclusively unified by a single
> agency? (12)

The possibilities inherent in the motion picture thus lend themselves to
"overexaggeration" on the part of many critics, as Marion Sheridan as-
tutely realized. In a 1936 essay ("Rescuing Civilization through Motion
Pictures") that adumbrates and undermines the criticisms of the Frankfurt School in "The Dialectic of Enlightenment," Sheridan suggests that
the salvation of the world will not be effected solely through didactic
pictures:

> The didactic has its place. It has not had the first rank in the fine arts;
> and it is with the fine arts that the future of the motion picture would
> seem to be. Recognition of the motion picture as a fine art with a future
> would tend to end the underestimation of the general intelligence. It
> would seem to be a way of raising the populace a point or two higher
> on the stages of culture. (170–71)

It is, of course, one of the dicta inspiring and animating the film appreciation movement that with this cultural evolution must come moral
evolution. "Rescuing civilization through motion pictures" sounds melodramatic to us; in the context of the reformist thought of its day, it appears to have sounded merely forward-looking.

The Character Education Movement

Film appreciation, in other words, was intimately connected with character education, which indeed attracted many of the same adherents. Indeed, a number of the early comments on the character education series "Secrets of Success" suggest that some people found character education interchangeable with film appreciation. For example, John H. Keene argued that "Secrets of Success" "might lead to a practice of analyzing all photoplays they see—which is excellent in itself" (quoted in LeSourd, "Plans," 6), suggesting a behavior that resembles one aim of film appreciation. Even so, character education films were possibly more ambitious than a film appreciation program; the "Secrets of Success" films, for example, were meant to be "effective conditioners of conduct" ("Announce," 6). More specifically, it was the hope of those developing the project that the films would promote particular character traits in students: "Each [film] is a one-reel cutting from a full-length feature film and each emphasizes the desirability of some fundamental trait, such as honesty, democracy, unselfishness, or temperance" ("Social Workers," 1).[22]

The character education movement appeared to its contemporaries to have begun with religious filmmaking, which was perceived to have failed through its very obviousness of purpose. As *The Motion Picture and the Family* reported,

> Almost from the time that motion pictures began their flickering portrayal of scenes and stories, far-visioned men have predicted their use in character education. Individuals and groups have proven their faith by investing large sums of money in an attempt to provide films for this purpose. The religious emphasis in these early character films was dominant and their contribution was significant, not so much in the attainment of their goals as in giving new directions for further efforts. ("Announce," 1)

Religious filmmaking was perhaps insufficiently modern and lively to attract a popular audience, which required "pictures that taught a lesson, yet were virile enough to appeal to red-blooded boys and girls" ("Announce," 1). Thus the "Secrets of Success" series, which provided re-edited versions of popular Hollywood product in snippets ranging from twenty to forty minutes, had the advantage of securing the attention of young people with the kinds of films that they would have attended anyway, albeit in a rather different setting (Jacobs, 36). "Secrets of Success" promised to restore control to precisely those figures—teachers, clergymen, and social workers—whose authority was waning under the influence of mass media.

This religious matrix is apparent in the membership of the committee that began the project. Evidently "Secrets of Success" was the result of a 1929 Public Relations Conference of the Motion Picture Industry in New York, which commissioned a committee on "Social Values in Motion Pictures" staffed by "representative churchmen" under the chairmanship of Howard M. LeSourd, dean of the graduate school of Boston University ("Announce," 1). Like many similar organizations, the committee stressed the generally ecumenical outlook of the participants, along with the film industry's desire to cooperate (Moley, *Hays Office,* 154–55).

Ecumenism needed emphasis because the interests of the very loose coalition of reform elements behind this project would not have profited if one set of suspect national values (Hollywood's) merely gave way to another equally general and equally remote from immediate control and application. One of the prized aspects of "Secrets of Success," therefore, was the power over the text promised to group leaders. Among the virtues promised in the introductory article in *The Motion Picture and the Family* was the series' "adaptability—Character building agencies can find in these pictures basic material for programs adapted to various types of communities and to all age groups" ("Announce," 6). This flexibility evidently addressed concerns arising from the perception that Hollywood had failed to resolve local community dissatisfaction with the quality of its product.

But the tension between a national standard of success and its local application informed the inner workings of the character education movement itself. The student experiencing the "Secrets of Success" series would find general release films being retooled for his or her particular moral education, so that the mass-produced products of Hollywood were actually fitted to consumers according to local mores, on a classroom-by-classroom basis. Even so, the desire for local responsiveness and adaptability on the one hand, and for uniform standards and effects on the other, did not always mesh smoothly.

While the administration of "Secrets of Success," for example, was designed to be local, the Committee on Social Values in Motion Pictures would only cooperate with those social service agencies and schools

willing to use the programs according to instructions and simultaneously willing to gather information about the series' efficacy. Significantly, using the "Secrets of Success" series required that discussion
leaders amass information for the benefit of the Committee on Social
Values through the completion of "report blanks" and the making out of
"detailed reports of the class discussions which were held after the showing of each film" ("Pastor," 3). Here the tension between national scheme
and local application grew most acute, inasmuch as when the program
became more sophisticated it also became more centralized, relying, for
example, on manuals, discussion outlines, and posters, all designed to
secure the proper use of the films, and, inevitably, restricting their range
of interpretations ("Plans," 6). The idea of a uniform national plan for
the use of these shortened versions of Hollywood films was also con-
templated as curricula became more standardized. George F. Zook approvingly commented to the committee that "we feel sure that the type
of use you are making of the films would fit into any national educational film plan that may be developed" (quoted in "Much Praise," 2).

Further, there appears to have been a certain separation between the
rank-and-file participation within individual schools or agencies and
the persons charged with developing and guiding the series. Frederic
Thrasher, one of the series' most effusive supporters, remarks about social science research into media that it is best gathered into the hands of
the experts who know what to look for and how to tabulate the data:
"Personal observation is valueless because, unless it is controlled by research techniques, it may as well give magic formulae as scientific results"
(8). Indeed, Thrasher speaks as a technocrat searching for the right *technical* solution to address social evils:

> One of the outstanding causes of the confusion, the conflict, the mal-
> adjustment, the distress, and the tragedy which we see all about us,
> is that science has not been applied to the solution of social and
> educational problems. . . . Social organization in America is chaotic
> and permeated with social maladjustments expressing themselves in
> poverty, unemployment, crime, political corruption, vice, and violence.
> Civilization itself is constantly threatened with the slow disintegration
> wrought by its own social weaknesses or the violent destruction brought
> about by war or revolution. (8)

While Thrasher both praises and laments the products of the machine age, he seems unaware of the irony of applying the remedy of tech-

nology to the problem of the social dislocation produced by technology. For Thrasher, it is obvious that no one has systematically applied technology to the problems of social control. He echoes Taylor, saying, "Science, for example, has developed unthought-of instrumentalities which may be turned to the uses of peace and social efficiency" (11), which smacks of Taylor's desire to see labor and management exist in harmony (Banta, 93). Whatever Thrasher's attitudes toward labor or management, he certainly endorsed the notion that greater efficiency reduced social friction. As we have seen, Dale too subscribed to such a view, when he urged greater reliance on intellect than on emotion.

Obviously, character education had the potential to reduce social friction insofar as it was designed to create good citizens efficiently and to a standard pattern. Information about the program was readily disseminated in education courses designed for teachers and social workers, such as one taught by Thrasher at New York University in the 1930s. Thrasher described his constituency as "composed of school teachers, superintendents of schools and others" (letter of 16 March 1933, National Board of Review Papers, Box 45). LeSourd actually made a presentation to Thrasher's class in 1935 with selected "Secrets of Success" films in tow (letter of 9 January 1935, National Board of Review Papers, Box 45). Notwithstanding his general confidence in standardized, technical solutions notwithstanding, Thrasher was certainly aware that a one-size-fits-all approach might not work with the varied cultures of the Lower West Side. In the process of drafting "A Proposed Community Motion Picture Program" (which would include use of "Secrets of Success") for the Lower West Side Motion Picture Council, the idea of local adaptability remained paramount, as indicated by minutes of meetings held by Thrasher and members of organizations including the National Board of Review. Thrasher ratified a wording suggestion from a colleague by saying, "Why not put it, Because of grave dangers of standardization in communities where conditions are different, it is recognized that this program will have to be adapted to local conditions" (minutes of the meeting of the Motion Picture Council, National Board of Review Papers, 21 December 1934, Box 20, "Children and the Motion Pictures" folders, 15).

To appeal to a variety of localities, the "character building agencies" were to function with all the attractiveness of the mass media. Careful avoidance of any obvious markers of uplift, just as in film appreciation, was one strategy that LeSourd particularly emphasized; he noted that "we of the Committee realized that if pictures were to be effective conditioners of conduct, they couldn't preach and the audience should be led to discover for themselves the moral in the situation portrayed" ("Films," 4). Social workers, too, were conscious that the touch had to be light. Frank Astor of the Department of Child Guidance of the New York City Public Schools noted that "if a teacher is formal, she can spoil it" (minutes of meeting of the Motion Picture Council, 21 December 1934, National Board of Review Papers, Box 20, "Children and the Motion Pictures" folders, 7), while a discussion outline for a renamed version of "Secrets of Success" called "Let's Talk It Over" cautioned discussion leaders against "any preaching or moralizing during the discussion or at the close" (Hays Papers, box 47, folder for 16–31 December 1934).

Nevertheless, those engaged in the project of character education planned to have a significant effect on the psyches of their charges. Both film appreciation and character education assume, as Jacobs notes, that film is an exceedingly stimulating medium. Some contemporary psychologists evidently thought that, given such stimulus as motion pictures represented, children must inevitably "work off" the effects in a constructive way or come to harm by bottling up the emotions thus aroused, since these children might later find themselves working off the stimulus in some other, less beneficial way.[23]

For example, Dr. F. A. Magoun praised "Secrets of Success" in these terms, quoted in "Noted Teacher":

> Psychologists have long maintained that the principal basis for disapproving the theater and the movies is that one's emotions are often aroused to fever pitch with no outlet. This stimulation without accompanying action is weakening to the will.
>
> But here we find both stimulation and accompanying action. After the class has seen the film, their aroused minds instantly find expression by contributing to the vitality of the discussion which follows. Nor is this all. The vigorous examination of a good idea is one of the surest ways of introducing it into one's actions.
>
> Thus the use of these films follows the best principles of pedagogy. First the student's interest is thoroughly aroused. Then his desire to develop ideas, instead of listening to just another lecture, is given outlet as well as stimulation. Then, having arrived at conclusions for himself, instead of having been forced to swallow predigested pap to which he will give little better than lip service, he is both ready and eager to try

out these conclusions in his own conduct of life. The kind of experience which develops us most is always that which increases our motor reactions. (6)

These insights appear to harmonize with Dewey's notion that education does not arise from contact with stimulus alone, but requires the intervention of culture: "Nature must indeed furnish its physical stimuli of light, sound, heat, etc., but the significance attaching to these, the interpretation made of them, depends upon the ways in which the society in which the child lives acts and reacts in reference to them" (*School*, 91). Yet Magoun's comment is ambiguous, suggesting on the one hand a close connection between emotional and even intellectual function and motor reflexes, which suggests a fairly deterministic view of human behavior, and on the other the possibility that a cultural authority could mold this behavior. Culture for Dewey, however, defines the sorts of responses appropriate to various stimuli; consequently, he has less faith, as Joas argues, that physiology is the sole agent determining human behavior.[24]

Dewey's views on art in fact suggest that he thought that arousal was a necessary precursor to interpretation, not an undesirable by-product of certain types of artistic practice:

> Perception is an act of the going-out of energy in order to receive, not a withholding of energy. To steep ourselves in a subject-matter we have first to plunge into it. When we are only passive to a scene, it overwhelms us and, for lack of answering activity, we do not perceive that which bears us down. We must summon energy and pitch it at a responsive key in order to *take* in. (*Art,* 53)

Magoun appears to agree that arousal is essential for learning, even if he is too willing to connect learning to motor activity. At any rate, one need not try to avoid arousal, but rather, one must simply ensure its proper discharge, in Magoun's view—suggesting an entirely new conception of reformist hopes for film.

It might seem, given their willingness to condemn individual unworthy motion pictures, that some elements of the character education and film appreciation movements hoped to restrict access to film because they considered the medium too overpowering and brutalizing. But in fact we find in Magoun's remarks an interesting willingness, indeed a desire, to promote the use of film to engage young people in a contemplation of moral problems. Film as raw stimulus is the "nature" to the character education movement's "culture." Magoun would no doubt have abhorred the idea of dispensing with film altogether, because many educators had concluded that film was the way to accomplish socializing and civilizing goals that had proven difficult or impossible with other tools.

Magoun's remarks suggest that the model we should adopt for character education is one of a "stimulus and response" dynamic controlled by some authority figure such as a teacher or clergy member. What both film appreciation and character education most desired was that the classroom take over the control of meanings derived from film exposure. However, if film appreciation hoped to inoculate children against misguided appeals to sentiment, character education seemed readier to exploit sentiment in securing children's attention. *Skippy* (Norman Taurog, 1931), one of the films in the "Secrets of Success" series, met both blame and praise for its manipulation of affect; based on a popular comic strip, this picture follows Jackie Cooper as a middle-class boy, son of a health inspector, and his much poorer friend as they attempt to acquire a few dollars for a dog license. The *Motion Picture Monthly* reported the disgust felt by *The Presbyterian,* which dismissed the film as "utterly wrong in its general philosophy, and also present[ing] a sickly sentimentalism on the part of silly parents" (quoted in "'Clean,'" 6). Side by side with this criticism, however, appeared a laudatory account from the Houston *Press* that noted, "when 'Skippy' threw himself on the bed with his little heart seemingly broken, and when a little girl in front of me cried over the death of a dog in the picture, I was not ashamed that I had to wipe something from my eyes also" (6).

The issue within character education, in other words, is not the avoidance of arousal but rather the use of arousal for other ends, social or intellectual. Indeed, the doyen of the character education movement, LeSourd, makes clear that previous attempts failed to work because of insufficiently compelling instructors:

> Motion pictures will become invaluable to the program of character education, because of their effectiveness in dealing with the emotions which are the essence of the dynamic factors of life. The general popularity of motion pictures arises almost entirely out of their emotional appeal. This constitutes at once their possible danger and their unlimited potentialities. Teachers vary greatly in their abilities to arouse proper feelings but when a film is made its effectiveness is uniform and universal. ("Use," 4)

Not only is the film more charismatic than any but the best teacher, but LeSourd's contemporaries appear to have read the standardized and potentially standardizing nature of film instruction within the film appreciation and character education movements as a distinct advance in the democratization of education.

Lewin, for example, also concerned with the relationship between efficient education and democracy, argues repeatedly that using films in classrooms is inherently more "democratic" along exactly the lines here presented by LeSourd: film eliminates the inequality of educational experience that results from poorly prepared or uncharismatic instructors and thus ensures that all children have access to the same standard instruction ("Experiments," 42; "Photoplays," 453; "Teachers," 296). Lewin argues that the film is a more efficient method of instruction precisely because it is more compelling than all but the best teachers, who "succeed through extraordinary traits of personality. They win and hold pupil-attention with ease. They glow with a continual magnetism that arouses real interest and compels love and admiration on the part of children" ("Photoplays," 453). From this description, we gather that the good teacher is a hero of the personality age, and so is the film that may come to assume a place at his or her side, which is exactly why film is so valuable in character formation.

But overreliance upon standardization troubles even Lewin, democratic rhetoric notwithstanding. On the one hand, we have the authoritative teacher made still more compelling through the use of a standardized tool, such as a series of films. On the other, we confront the possibility that teachers may not be necessary in the brave new world of machine-assisted visual education. Lewin hastens to dismiss the second option by commenting that "all the statistics and standardized tests in the world cannot take the place of the true teacher, and all the electrical devices in the world, whether silent or talking, are of little value without the personal touch and the guiding hand of the teacher" ("Progress," 9). This view has two important ramifications. The first is that the teacher is indispensable in his or her role of tailoring the standardized film experience to the local circumstances. The second is that the teacher's growing facility with visual education generally suggests a new mandate in intervening in children's consumption of the products of the mass media.

The emphasis invariably falls on classroom activity, which seeks to guide students' response to a stimulus provided by Hollywood. Activity is also the answering dynamism that permits students to work off any excess stimulation, even as it undoes the passive attitudes that commercial leisure fosters. For example, the short clips constituting the individual films of the "Secrets of Success" series often omit the resolution of the moral problem shown, on the principle that the students themselves should supply it, under the guidance of their leader, as Deweyan pedagogy urges.

Even if the script for working through the "Secrets of Success" films tended to standardize the teacher's role through reliance on "discussion outlines," the actual discussion had the potential to escape the script, because students were evidently accustomed to debating and voting on issues. In an undated "Stenographic Report of Character Education Lesson" taking place in Bronx Public School 82, we see some ways of working out the concerns raised by *Sooky* (Norman Taurog, 1931), a sequel to *Skippy* in which a boy's father is taken away to the poorhouse while Skippy hopes to forestall the removal of his chum by charging groceries for the child without his own father's permission. Children appear to be accustomed to voting on issues (sometimes by a show of hands), appointing groups for debate, and forming committees to report back to the larger class. The stenographic record does not report the results of activities beyond the show of hands, but the children offer all of them as ways of addressing the moral questions raised, possibly reflecting a training in democracy (or group-process dynamics) they were accustomed to using in other subject areas. When the teacher asks, "After seeing this picture how many think you would know what you would do?" in such moral dilemmas, the answers range from the personal, "I would consider my parents," to the sociological and anti-Hooverite, "I think instead of spending money for an institution they should clean up Shantytown and spend their money in giving the people food and in giving them shelter also" (National Board of Review Papers, Box 18, "Bureau of Visual Education" folder).

Voting, debating, appointing committees, imagining oneself in the roles of the characters were activities that clearly agreed, as well, with those the film appreciation program promoted. Within the Deweyan context, efforts to engage students emotionally and even physically appeared valuable: they undercut the passivity encouraged by mass culture; they suggested that film texts always led back to a reality or work outside the film itself; and they allowed students to work off the arousal generated by the film text, in a supervised fashion. But film was clearly indispensable in this regime because it *did* arouse students. Here the rhetoric of self-commodification undergoes an interesting twist, as students are not being urged to model themselves on the heroes of the personality age so much as teachers are being urged to harness these heroes for moral instruction.

The emphasis on dynamism in LeSourd's comments clearly marks character education as resulting in part from an internalization of the rhetoric of the culture of personality. As LeSourd notes of the title of the series, "We call these pictures *Secrets of Success,* although no one seems enthusiastic about the name. Youngsters like secrets; they are interested in success. We did not wish to suggest to groups seeing the pictures any effort to improve them" ("Films," 5). The very desire to avoid obvious uplift maneuvers the language of character education into the get-ahead rhetoric of correspondence schools, suggesting an important area of overlap between the two self-fashioning projects.

Conclusion

Along similar lines, Lears has documented the incursions of advertising into zones previously under the sway of education, quoting a J. Walter Thompson newsletter of 1916 that "one of the very definite phases of our work is to be educators" (*Fables,* 206). Graebner argues in a similar context that group-process psychology came to occupy the sphere formerly controlled by organized religion partly through the hopes for social improvement inspired by the social sciences (27). Educators might have deplored the easy promises made by advertisers, but they now found themselves borrowing advertisers' terms and frames of reference. Moreover, the two disciplines shared at least one methodology in their deployment of group-process psychology; its practical application in both education and advertising smacked more of promoting individual adjustment to the status quo than of prizing inquiry and direct experience, with all the possibilities for change that Dewey and James saw in the latter.

One way in which the educational establishment lost the moral high ground vis-à-vis advertising was by attempting to teach the "proper use of leisure time." While the endorsement or criticism of individual films appears a reasonable activity in this context, especially given the Progressive hopes for cinema as an institution, in practice such endorsements mandated a certain uncomfortable closeness between education and advertising. As early as 1911, an advertising manager for Wanamaker's Department Store argued that "advertising is not to sell, but to help people to buy. . . . We stand in the shoes of the customer. We are outside, not behind the counter. We are counselors for the public" (quoted in Lears, *Fables*, 205), a statement that could serve as the mantra of the better-films movement and its educational offshoots. The project of molding consumer behavior within the schools was the very project that tended to undermine the ideals of the Deweyan curriculum. Molding taste often turned out to be a prescriptive enterprise—it assumed the correct answer and then established the means of arriving at it, rather than promoting a mode of inquiry that would permit students to develop their own standards. Some educators did attempt to analyze students' standards—and discovered, significantly, that their subjects already watched films with reasonable discrimination and self-awareness.[25] Nonetheless, the film education movement at times assumed a gap between the standards revealed in movies and the standards that governed American culture at large, an assumption possibly not warranted.

Dewey was himself approached in connection with a Payne Fund study that aimed to establish proper standards for motion pictures. His response was to suggest a descriptive enterprise before the Payne Fund undertook the prescriptive one: "He doubts whether any compilation of standards, made in advance, and without direct relation to an analysis of the films, would fit the need very closely," William Short of the Payne Fund summarized Dewey's answer (Jowett, Jarvie, and Fuller, 76). Dewey's view that "society's values were in too great a state of flux for one standard to be defined for all viewers and that moral values should be considered in the contexts of particular communities or groups" no doubt resulted from his conviction that the meaning of most human activities was dependent on their contexts (Jowett, Jarvie, and Fuller, 76), which was not the attitude of those who felt that film was perverting American taste. And while Dewey felt that the teacher's authority was necessary in the classroom to guide the progress of inquiry, he never intended that authority to *substitute* for inquiry.

To the extent that the educational establishment wished to use the new media as a stalking horse for its own agenda, it parted company from the Deweyan ideal. But, as this chapter has sought to demonstrate, the very changes in the American curriculum that appeared to stem from Dewey's pedagogy materially assisted the process of mutual colonization between the film industry and the educational establishment. Education felt that it too must partake of the remedies of the personality age by becoming more relevant, more attractive, and more compelling—all advantages that the incorporation of film into the curriculum could secure. At the same time, the film industry perceived the advantages, especially at the local level, to be derived from its affiliation with education in terms of the best possible product endorsement to be had. And both the film industry and the educational establishment wrestled with the problem of authority in a democracy; both deployed the rhetoric of the plebiscite at the box office to justify their relationships to their constituents, and each attempted to reconcile the desire for centralized control with the acknowledgment that in a democracy such authority was, ideally if not actually, dispersed to the people.

Thus both film appreciation and character education were alternatives to the approach to controlling film content enshrined in the enforcement of the Production Code (discussed in chapter 4). Film appreciation in particular hoped to ratify the sovereign individual conscience, capable of making informed decisions about film attendance. In practice, the ideologues of the film appreciation movement often had covert, specific standards about the kinds of films they hoped children would come to prefer; even so, they rejected the solution of controlling film content at the source, citing the unhappy political implications of such prior restraint of expression. Character education, often promoted through the film appreciation curriculum or by parallel structures, was in many respects much more ambitious than was the Production Code. The more forceful voices speaking on behalf of the Catholic Church (such as Joseph Breen and Martin Quigley) were satisfied when film did not preach error by flouting "natural law," and indeed preferred harmless entertainment to "message" films. Character education, on the other hand, hoped to harness the inherently persuasive and alluring qualities of film to mold children's characters through a form of automated group-process dynamic. Character education, in other words, was a later manifestation of the secularization of Protestant aims described in chapter 4, and one, significantly, that did not disappear even when the Production Code established the Catholic approach as temporarily dominant.

Conclusion

It goes without saying that the film industry was a major site of social discomfort with regard to the vexing question of consumption in a modern democracy. Yet one of the issues that this book has explored is how a discomfort engendered by ever more efficient consumption is at least partially linked to an understanding that conventional definitions of authorship were troubled not only by the industrial and professional standards of the film industry, which appeared to mechanize literary endeavor, but also by notions of a public sovereign at the box office, which unseated any conception of a unified, identifiable, and controllable author. Here we confront the central paradox of this study: each of the four cases discussed suggests that the public has a place in filmmaking, just as each ratifies some form of the notion that the public is the ultimate source of film texts. Yet from 1916 to 1934 we see, to generalize, a constriction in the definition of means of public access to the film industry, from hands-on filmmaking in the case of the juvenile series fiction and the Palmer Photoplay Corporation to the more circumscribed aim of creating an educated filmgoer, an aim that begins to dominate the rhetoric of the denominational and educational incursions into Hollywood. What is particularly striking here is that the outsider *institutions*, namely churches and the educational establishment, offered the public perhaps the narrowest definition of participation in the film industry (a definition serving their own agendas) at more or less the same moment (beginning in the 1920s) that the film industry placed increasing obstacles before members of the public, such as the development of such phenomena as Central Casting, which was designed to dim the ardor of small-town would-be actors and actresses hoping to strike it big in Hollywood.

What one sees here is not a conspiracy of reformers working hand in glove with the film industry to exclude the public from meaningful participation but rather an extremely complicated struggle for control over textual production waged simultaneously by Hollywood, a large variety of reformers, and some members of the public. Hollywood and the forces of reform increasingly defined participation as authority at the box office rather than behind the camera, because to do so served the independent centralizing tendencies of both. Reform pressure on Hollywood may have relegated the public to more passive roles, inasmuch as it urged the installation of structures, such as the Production Code, that would eliminate "vulgarity," which at times may have operated as a code word for certain kinds of popular participation. If the Palmer Photoplay Corporation, for example, promised subscribers the means of telling a personal story, it nonetheless offered to organize that story in terms of Hollywood's own professional methods, and simultaneously served to disseminate the Production Code's precursors to a wider audience by informing would-be scenarists of the "Don'ts and Be Carefuls."

Hollywood's increasingly technocratic organization, in other words, arose not only from the exigencies of its business methods but also in response to a climate of reform that recognized such organization as more professional and therefore more reliable, a concept of good business practice already latent in the earliest case taken up here, the juvenile series fiction. The Depression gave new urgency to a definition of the film industry as responsive to the public weal even as it suggested that the industry's organizing structures (including some of those contributed by reformers) were inadequate. As the discussion in chapter 4 has shown, reformers did not envision direct access to the film industry in the interests of individual social rise, as did Palmer Photoplay and the series fiction, but rather as organized pressure at the box office.

The pattern that emerges from chapters 4 and 5 suggests that when Hollywood extended its reach and threatened local mores through ever more efficient means of film production, reformers responded by attempting to create totalizing structures for the control of film reception, either by flirting with a film distribution system outside the mainstream or by attempting to assert themselves as authorities at presenting films to audiences in the name of elevating public taste. Again, Hollywood did not necessarily resist these incursions but instead at times sought to capitalize upon them, using, for example, the National Council of Teachers of English study guides to leverage the marketing of worthy but less popular films.

The preferred rhetorics of all four cases explored in this book purport to offer something to the public. The first two promise to purvey the secret of power *within* the film industry; the second two promise to purvey power *over* the film industry. In all instances the road to power involves the mastery of a variety of extratextual knowledges, whether through explicating Hollywood's professional standards or by revealing the economic or social structures of the industry in order to mobilize the public for a particular purpose. Clearly the concept of public as author requires the circulation of this extratextual knowledge, and suggests that a rich field for pursuing the ramifications of this book may lie in an examination of the relationship between the cases studied here and, for example, contemporaneous fan discourse, which was probably the most significant extratextual knowledge at work during this period.

But as this conclusion seeks to illustrate, another inquiry that would repay investigation would be the project of tracing the phenomena examined in this study beyond the chronological bounds in place here. While the installation of the Production Code brought about a moment of equipoise, lasting at least until the late 1940s, there is no doubt that the desires and intentions expressed by individuals and groups in this investigation are still at large in the American public's relationship with its entertainment industries. Naturally, these relationships have changed significantly, in part because the institutional backdrop has changed enormously. For example, any account of the further fortunes of these approaches to individual or group participation in the film industry would have to take into account the rise of television and the increasing numbers of cable and community access programming outlets.

Additionally, and perhaps still more importantly, we must note the proliferation of personal video cameras and the development of the internet not only as a means of establishing and connecting groups of like-minded individuals but also as a nascent exhibition medium for homemade or independent video work. Whereas the rhetorics examined in the body of the present study all suggest the presence of substantial barriers between the amateur and the professional, albeit barriers that may be breached by the hard-working and talented, more recently the world of amateur filmmaking has become increasingly rich and strange. Arguably, there is less sense today that angels with flaming swords are barring the way to the paradise of individual or group self-expression.

Indeed, this sense that the old gatekeepers have been swept away by the significant changes in the structure of the industry over the past fifty years is perhaps the largest single alteration in the rhetorics of public participation in film since the conclusion of the period under study here. Film studies, for example, have triumphantly entered the academy, and now constitute a well-established subject within the humanities, with the result that they have become less useful in settings that hope to coopt them as the means for, say, character education. They are now no more likely to be used for that purpose than is literature, precisely because film studies have now been thoroughly absorbed into the analytical categories that govern academic discourse even at the high school level. Series fiction about filmmaking has likewise vanished, not because series fiction is currently in desuetude, but because we inhabit a technological and entrepreneurial culture so thoroughly acculturated to the idea of success at a young age that these particular bildungsromane no longer have the power to thrill as they did in the 1910s and 1920s. We expect children to be among the most savvy users of any technology, as we expect them to be among the most dedicated consumers and critics of popular culture, which has itself become technological in form and content as never before.

Observing that the old gatekeepers have become scarce is not to say, however, that lineal descendants of the impulses, groups, and rhetorics explored in this book are not still at large. Although film appreciation may be in abeyance, for example, character education through film has its present-day adherents. Take the case of the Ghetto Film School in the South Bronx, founded in 2000. A program designed to provide inner-city youth with a means of self-expression that will endow them simultaneously with possibly marketable skills and with self-esteem, the school emphasizes self-expression above didacticism but is nonetheless calculated to turn at-risk teenagers into articulate and law-abiding members of society. "It's not just about taking the students and trying to do social work," notes a college student who taught in the program during the summer of 2001 (quoted in Theim), but clearly "social work" is, as with the students involved in character education in the 1930s, a significant part of the agenda.

We may also see a descendent of character education in some activities of the management philosophy known as Appreciative Inquiry or AI, which itself resembles the group-process psychology discussed in chapter 5. Joe Hall, both the founder of the Ghetto Film School and a doyen of AI, describes in "What Is Appreciative Inquiry?" his experiences with a South Bronx neighborhood organization, experiences that, he notes, included a group discussion of the film *Good Will Hunting*. Will, the film's protagonist, emerges within this discussion as a suitable role model for residents of Kelly Street because of his ultimate commitment to making a difference in his hometown despite the community's various unsatisfactory aspects. Just as character educators in the 1930s required high school students to contemplate the moral successes and failures of the protagonists of selected films in order to strengthen their own moral capacities, the AI facilitators for the "Banana Kelly" group— which, as Hall notes, focuses particularly on young people—evidently have found that guided discussions of suitable films are important to the program's goal of "building quality people" (Hall and Hammond).

As for nonacademic instruction in screenwriting, consider a computer program called StoryCraft, which offers to automate precisely the functions the Palmer materials were consolidating in print-based ways. Thus StoryCraft's promotional material notes that the program "establishes the ideal category for your story," "matches your story with a time-proven pattern[,] and instantly outlines your story" using one of eighteen (as opposed to the Palmer *Photoplay Plot Encyclopedia*'s thirty-six) "basic patterns of stories that all the great myths and literary classics fall under." In diction redolent of generations of advertising slogans, StoryCraft "banishes writer's block by simplifying the actual writing process" ("StoryCraft," 1). StoryCraft also relies upon the charm of wedding technology to the precepts of "history's greatest writing coaches" (3) and bills itself as applicable to a wide range of writing, from screenplays to short stories, just as Palmer's later material offered to do. And, like the Palmer material, StoryCraft offers to both free and discipline the user—brainstorming becomes easier and excuses (owing to the many automatic features and the rigorous system) become more difficult. Indeed, the only elements of Palmer's advertising missing here are those that address themselves explicitly and exclusively to the transformation of personality attendant upon ultimate success in the business of writing screenplays, perhaps because that message has been so thoroughly absorbed by the American public that it does not bear repeating here.

The idea of rapid rise to a position of prominence in the entertainment industry through the provision of just the right personal story, however, is still very much with us. There are pages and pages of information available on the internet for budding scenario writers eager to send screenplays off to contests. Homemade documentaries and even fiction films find homes on both the internet and a variety of cable channels. The amateur filmmaker, however described, has more outlets for his or her product than ever, and, more importantly, even established film culture over the past thirty years has become exceptionally welcoming to the outsider. In other words, our present film culture is one in which the outsider has a permanent and well-regarded place. Independent filmmakers are prized even within the industry.

Something like ministerial filmmaking is also back. Organized, if not militant, Christian entrants into the field of popular culture are more visible than in decades, and they are succeeding in the marketplace as never before. As a July 2001 *Newsweek* cover story reports, this success is both desirable and undesirable in the eyes of its sternest critics within the evangelizing movement. As always, an accessible treatment of the Christian message offers the hope of evangelizing on a scale hitherto only dreamed of. At the same time, of course, such a mission always courts two possible disadvantages, pointed out by *Newsweek* (see Ali, 43). The first is the consequences of adopting the methods of the "enemy": is rock music, for example, so decadent that there can be no such thing as a doctrinally appropriate rock song of praise, or a prayer set to hip-hop rhythm? The second problem is the issue of commodification: we now have not only Christian records, t-shirts, and videos, but also superheroes, such as "Bibleman," that can be marketed like any other proprietary figure in the vast sea of popular culture. While this marketing ensures wide exposure, it may simultaneously require certain modifications of the central message, so that dogma is softened or obscured to appeal to as many people as possible.

Significantly, theatrical film production has been the weakest sector of this market overall, no doubt as a consequence of the expense and risks involved. (Film production for release direct to video has been more lucrative, on the whole.) Yet we now encounter the phenomenon of what might be called, for want of a better term, the evangelical action film, in the form of Matthew Crouch's *The Omega Code* (1999) and its sequel, *Megiddo* (2001). These films have received significant theatrical releases and have done reasonably well at the box office—*The Omega Code* proved to be the third highest-grossing independent film of 1999, with earnings of over $13 million in its first year of release ("Christian Film"). Significantly, the film deals with the efforts of a motivational speaker who converts to Christianity to thwart a satanic media kingpin's plans for world domination; in other words, the plot hinges on the desirability of a comparative outsider toppling an insider who is using his power and authority in inappropriate ways, so that this production may be read as an allegory, albeit presumably unintentional, for the kinds of 1920s religious intervention in film detailed in chapter four. That the "omega code" in question is biblical in origin and has been wrongfully coopted by the kingpin suggests that the happy ending entails less a complete redistribution of power than its restoration to its rightful stewards—who are members of the middle class rather than tycoons.

The success of films such as *The Omega Code* need not be considered a fluke. Christian Filmmaker Ministries, for example, is a sophisticated networking effort designed to assist right-thinking small independent producers through the maintenance of a clearinghouse for information and possible funding. The organization's mission statement, posted on its website, declares that the Ministries' main objective is to "Go ye into all the world and preach the Gospel . . . in a creative and entertaining NEW way, producing films and videos with excellent quality so we may reach the lost through the power of Motion Picture Films and Television." This approach is pouring old wine into new bottles with a vengeance. Ambitions notwithstanding, funding appears to be in short supply, since the website notes that the money to fund a grants program serving Christian filmmakers is still being prayed for. Crouch's recent success, on the other hand, suggests that capital might be forthcoming for the right kind of sufficiently ambitious filmmaking. I might add that such films dovetail with their contemporary mainstream Hollywood products better than might be expected; the typical Arnold Schwarzenegger release, say, shares *The Omega Code*'s concern with representations of apocalypse, albeit in a somewhat different context. Clearly, Crouch and other contemporary film producers want their products to demonstrate two things as far as the packaging of content is concerned: mastery of the present-day idioms of popular filmmaking (hence the action/adventure paradigm) and first-rate production values.

Unfortunately for even Christian filmmakers, the problem of appropriate affect remains. As dramatic in terms of box office success and special effects as it was, *The Omega Code* provoked criticism in some Christian circles precisely because it appeared to be attempting to "evangeliz[e] through fear" (Bourke), as both Barbara Nicolosi, a Christian screenwriter, and Bobby Downes, an ex-missionary and the producer of *Mercy Streets*, note. Downes even invokes the language of psychological experiments when he observes that "it's wrong to use fear and what's really wrong is to make people sit there in a theater and feel cornered" (quoted in Bourke). Their ideal of Christian filmmaking is something less splashy and emotionally aggressive. Journalist Philippa Bourke quotes Nicolosi as remarking, "I'd like to see us just make movies about our own experience. . . . If I as a believer can, in a beautiful way, show my experience of faith, that will be compelling for you to behold. You will be touched." The comment is reminiscent of the discourse both of earlier religious and educational reformers, who recognized (for all their distrust of inappropriate affect) that affect on a small scale might be powerful for good, and of the manuals published by the Palmer Photoplay Corporation, which consistently reminded students that their individual life experience might be widely marketable.

Nicolosi's experience resonates with Palmer Photoplay in other respects. She is the founder of Act One Writing for Hollywood, a Christian screenwriting school begun in 1999 with the hopes of training a cadre of able Christian photoplaywrights. Like Frederick Palmer, Nicolosi, who has substantial Hollywood credits of many kinds, operates here as an insider willing to explain the mysteries of the industry to those outside it. That Act One's faculty can claim experience in children's television, project development, film and television consulting, and production, among other areas of the business (Bourke), presumably suggests to potential students that if they can absorb enough of their mentors' advice, they will emerge from the program with the requisite knowledge to enter a profession by no means easy of access. We may well recall here that the credentials of the Palmer Photoplay personnel and lengthy discussions of Hollywood's elaborate gate-keeping apparatus were used to attract subscribers to the Palmer System. The difference is that what animates these new outsiders is not the desire to assimilate their followers to the glamour and power of the industry, but rather to assimilate the industry to the glamour and power of Christianity.

But, of course, the religious efforts discussed in chapter 4 included not only filmmaking but also the attempt to exert control over Hollywood products. Similarly, we see today both a newly resurgent film production and some pressure at the box office. Even now that the economic evils of the classical Hollywood cinema, such as block booking and blind selling, have been banished, there remain tensions between local pressure groups and national distributors. The conflict, however, is more likely to be fought out at the video store than at the theater: retailers such as Wal-Mart and Blockbuster must select their wares for sale or rental with an eye to their local constituencies.

Certain films are objectionable enough to engender boycotts not only of individual titles but of all products associated with a particular manufacturer. Consider the form taken by organized protests over Antonia Bird's 1994 *Priest,* which depicts the travails of a homosexual priest who attempts to minister to an abused child. The first objections came from certain Catholic groups responding to what they took to be representations of tormented or misled priests who were dysfunctional owing to their Catholicism and not in spite of it, and took the form of public protest and the sale of Disney stock by such organizations as the Knights of Columbus. Matters then widened to involve Protestant groups who were less concerned with the portrayal of Catholic clerics than with the prospect of the extension of insurance benefits to employees of the Disney Corporation involved in same-sex partnerships. This boycott is strangely reminiscent of the star scandals that initiated the Hays Office in 1922, during which objections to the morals of stars finally encompassed concern over film content as well. Film now has First Amendment protections, of course, leaving such concerned citizens' groups with little recourse other than making their displeasure known in economic terms. Yet the rhetoric of the plebiscite at the box office remains to such protestors. As Crouch observes of his filmmaking project, "I truly believe that once the Christian community understands that they have a vote by buying a ticket, they will become the country's largest single market" (quoted in Peyser, 45).

The persistence of these rhetorics indicates that we expect our entertainment to be more than merely amusing. We assume or demand that it have the power to transform us, or, perhaps better put, our access to entertainment is a means of self-transformation, in which variously we can commodify our own experience to make money or even to win souls. Allied to this view is the equally persistent expectation that entertainment be somehow representative of the American public. Outsiders who wish to enter the market perennially justify their desire by claiming to speak for a significant segment of the population that may not yet have found its voice within the industry. By implication, entertainment without representation is tyranny; presumably, if there exist demographics as yet unserved, their spokespersons have not only the right but also the obligation to step forward to demand that appropriate stories be produced for them. The notion of market as plebiscite has been the great constant in American entertainment since at least the 1910s, even if the modes of enfranchisement have changed substantially.

Notes

1. The Rhetorics of Democracy

1. From the 1890s into the early years of this century, populism condemned disparities of power and wealth as antidemocratic; see Pells (126) and Hofstadter (64–66). I apply "populist" in this connection to describe disaffection with big business and "special interests"—in this case, Hollywood and the Hays Office—coupled with a distrust of centralized power, which, as portions of this chapter will detail, Hollywood certainly wielded. Hence "populist" seems aptly to describe the Legion of Decency campaign and the rhetoric that preceded it, which suggested that excessive power had been vested in a corrupt institution and that demonstrations of popular disaffection, such as withholding attendance at movie theaters, could achieve redress. Alan Brinkley particularly notes the continuing thread in populist rhetoric (as late as the time of Father Coughlin, discussed in chapter 4) concerned with the protection of local against national interests (164), an aim that united reformers of film across denominational lines.

2. In linking the term "democratic" to film, I follow Garth Jowett, who uses the word in this connection in his important 1976 book, *Film, the Democratic Art*. While Jowett suggests that film was early identified as a mass art form that awakened the concern and interest of reformers, he does not in fact advance a book-length argument about how competing definitions of democracy, or rhetorics of democracy potentially at odds with each other, shaped the use of the medium.

3. T. J. Jackson Lears argues that the low mental age revealed by the Army intelligence tests had strong repercussions within the advertising industry, where "emphasis on the stupidity of the audience comported well with the rhetoric of mass man, allowing advertising spokesmen to preserve their own pretensions to rationality and objectivity while engaging in deception of the multitudes" (*Fables*, 233).

4. Possibly naively, the National Board of Review stuck to its faith in audience education and asserted that the average mental age was increasing. As late as 1934, Frances Taylor Patterson asserted, "I think it is not too sanguine to say that the level of audience intelligence has been raised through the instrumentality of the review

groups of the National Board of Review. . . . If it used to be 14 years old, it is now at least 21" (9).

5. This view of the audience was one that Will Hays endorsed in his refusal to promote the adaptation of the "prevalent type" of book and play. For a fuller discussion of this issue, see Richard Maltby, "Book."

6. I thank Janet Staiger for drawing this reference to my attention.

7. Indeed, this question of marketing is where the field of scientific management was heading in that decade (Haber, 164), although Taylor's own preoccupations were purely with industrial production.

8. For further discussion of Barton, see Lears, "Salvation" (30–38), and Maltby, "King" (198–203). The latter essay argues that Cecil B. DeMille's *The King of Kings* is an adaptation of Barton's work, an insight that exposes the interconnection of religion, salesmanship, and film in the 1920s. *The Man Nobody Knows* was itself turned into a motion picture soon after its publication in book form (Lears, "Salvation," 34). Finally, Barton may have been following the example of Gerald Stanley Lee, whose 1913 book *Crowds* suggested "business and efficiency [were] the central concerns of Jesus" (Bush, 135).

9. Maltby has written extensively and very helpfully on this subject (see "Baby Face," "Code," and "Book").

2. Acting Naturally

1. For a comprehensive list of titles within each series, see Johnson, *Pseudonyms*. These series have attracted little sustained scholarly attention, although Carol Billman and Deidre Johnson have produced book-length studies of the Stratemeyer Syndicate. John Parris Springer, who concerns himself primarily with adult fictions about Hollywood, notes in passing that the technological books represent the first appearance of "fiction about moviemaking" (10).

2. By far the most common element of film technology to be explained in series fiction is editing. Virtually every early work contains an explanation such as the following: "In the case of trick pictures, or when some accident is shown, the camera takes views up to a certain point with real persons posing before it. Then the mechanism is stopped, 'dummies' are substituted for real personages, and the taking of the film goes on" (Hope, *Oak Farm*, 53–54).

3. As in much American children's fiction, orphanhood here serves as a means of involuntary emancipation. Ben Singer observes a similar emancipation on the part of heroines of serial melodramas, who exhibit "a variety of traditionally 'masculine' qualities: physical strength and endurance, self-reliance, courage, social authority, and freedom to explore novel experiences outside the domestic sphere" (91).

4. Carol Nackenoff observes that Alger almost always steers his characters away from factory work, suggesting that this particular aspect of series fiction as employment primer has a long lineage (81–85). Clearly, a common aim of Alger's and Stratemeyer's books is to identify the mode of employment that will assist characters to stay in or reach the genteel middle class.

5. The issue of domesticity is discussed at greater length in a somewhat different connection in Morey and Nelson (3–4).

6. Ina Rae Hark's survey of theater management manuals from the 1910s into the 1990s points up the consistent emphasis upon cleanliness and homelike atmo-

sphere as the goal for the good theater manager. She argues that the demands of running a theater as a kind of superior home sometimes conflict with the demands of twentieth-century business masculinity. The rhetorical disjunction that might otherwise have arisen here was typically handled, in both the manuals and the Chums series, by rendering good housekeeping a commodity, rather than by folding it into the personal attributes of the male theater manager (180).

7. *Aunt Jane's Nieces Out West* adds that the same is true of the didactic ("If we hurled righteousness at [viewers] they would soon desert us" [Baum, 25]), so that educational elements "must be liberally interspersed with scenes of action and human interest" (26), in effect turning a sermon into a work of art and proving that the box-office plebiscite is actually beneficial to Hollywood's product.

8. A similar moment occurs in another series signed by "Appleton" (a pseudonym sometimes used by Stratemeyer) when, in *The Motion Picture Chums' War Spectacle* (1916), Frank establishes contact with two young people by remarking, "I think I saw you folks in the picture show. . . . Did you like it?" (138). When the boy expresses his approval, Frank explains, "Two other fellows and I own it in partnership. We like to know when our patrons are pleased" (139).

9. Ruth Fielding, the only other female capitalist of any note in these filmmaking series, finds herself perennially at risk of invasion of privacy primarily because she is known to the public through her performances as an actress and her well-publicized success as a scenarist.

10. Because the Moving Picture Boys are more likely to be shooting documentaries or news features than screenplays, the object of the thefts that occur in this series is often the exposed film itself. In this case the malefactors are arguably making off with another type of "performance," namely the luck, skill, and steady nerves that have enabled the Boys to record earthquakes, volcanic eruptions, charging lions, and so on. Here again, the artistic aim is to combine high drama and reality to maximize the emotional effect for the audience, and the plot tension comes from the question of who should rightly receive credit for producing this effect.

11. The importance of Ruth's ability as "location woman" to the success of her company was previously established in *Ruth Fielding in the Far North, or The Lost Motion Picture Company* (1924), when her fiancé, Tom, "found what seemed to him a perfect setting for the film, but each time Ruth had some objection to offer. Tom, while puzzled, did not question her judgment. He was aware that, as far as the making of moving pictures was concerned, it was far superior to his" (87).

12. This contest is an excellent example of how the series deploys rather outdated or occasionally misleading information about the film industry. While such contests as Ruth enters here were common in the 1910s and early 1920s, they became rarer by the end of the decade, and, moreover, did not address established scenario writers but amateurs, which Ruth has long ceased to be by this point in her career.

3. Fashioning the Self to Fashion the Film

I thank Janet Staiger for graciously sharing her collection of Palmer Photoplay Corporation materials, and John Belton for his helpful suggestions in revising an earlier version of this chapter published in *Film History*.

1. The Cornell alumni office, however, can find no record of Palmer's having matriculated.

2. Sadly, I have been unable to locate a copy of this test, but Don Ryan, editor of *Don Ryan's Magazine* ("published at irreverent intervals") quotes the "dramatic perception" test from the questionnaire sent to him: "Crossing the desert with Buck Logan, a desperate prisoner, a western sheriff comes upon two children, unconscious and dying of thirst. Logan has made a futile attempt to escape but has had no chance to use the small automatic which he has concealed in his boot. They have but one horse, no water, and Logan cannot walk because of a broken leg. So the sheriff faces a tragic choice. The horse can carry Logan or the children, but not both Logan and the children. Which shall the sheriff save—Buck Logan or the kiddies?" (9). The hard-bitten Ryan suggests "that Buck shoot the horse. Then all the characters will die in the desert and the public will be spared" (10).

3. See, for example, Alfred Hustwick's proposals in "Motion Pictures—To-day and To-morrow: 5." Writing in 1926, Hustwick calls for more systematic study of motion picture manufacture, including the establishment of a research bureau in each studio and closer relations with universities and colleges.

4. Significantly, however, Frederick Palmer rebuts these charges. While generally praising the honor and intelligence of the editors (both male and female), he suggests that the gatekeepers "serve merely to eliminate the utter impossibilities, and under no circumstances do they have any part in the actual choosing of stories," thus sidestepping the larger issue of their competence (Palmer, "Perturbation," 590–91).

5. The Palmer house organ was variously called *The Photoplaywright* (1919–21), *The Photodramatist* and *Photodramatist* (1921–23), and *The Story World and Photodramatist* (1923 on). Starting in July 1921, the journal was the "official organ of the Screen Writers' Guild of the Authors' League of America" (LeBerthon, 13).

6. The first installment was $15.00 (*Little Stories*, 4). The prices are quoted in undated subscription brochures; I thank Richard Koszarski for sharing with me his collection of Palmer Photoplay material, and I thank John Belton for bringing those sources to my attention. The Palmer price, incidentally, compares unfavorably with the cost of a screenwriting course at Columbia, which was a mere $37.50 in 1938 (National Board of Review Papers, Box 21, "Columbia University" folder).

7. The Home Correspondence School also published Henry Albert Philips's analogue to *The Photoplay Plot Encyclopedia*, under the title *The Universal Plot Catalogue*, of which the major feature is an alphabetical index of plot subjects that runs from "abdication" to "zodiac." It emphasized efficient writing through the slogan "Valuable Plot Material mislaid becomes an obstacle—instead of an alliance—in plot building" (129). Needless to say, the concept of building a plot borrows from engineering metaphors, in keeping with the Taylorist discourse of efficiency.

8. On the importance of action, see, for instance, Palmer, "Whence Future Photoplays?" (189) and the *Palmer Plan Handbook*'s "What Is a Photoplay?" (1918, 12–15). For other lectures involving aspects of technique, see Beban and Clift.

9. Staiger argues that as melodrama moved into the twentieth century it was particularly likely to emphasize the necessity of the protagonist's eventual comprehension of the consequences of his or her agency and the relationship of that agency to other social structures (*Bad Women*, 81–83).

10. Cecelia Tichi has explored the applications of Taylor's thought in contemporary literature, where efficiency was the goal of expression. She notes that Taylorism

"redefined the artists' and writers' relation to their work in an age of assembly-line manufacturing from component parts. In the waning of the crafts era, it provided a new self-identification for the artist. One could now be a designer-engineer" (79). This new approach obviously had the effect of realigning definitions of talent and was possibly one of the respects in which Palmer's advertising could finesse the question of innate writerly gifts, viewed from an older, conventionally literary standpoint.

11. Since Palmer claimed to have sold some sixty thousand dollars worth of scenarios by this time (Tidden, 553), it is possible that contest prizes constituted a fairly high proportion of the money earned by students.

12. I am indebted to Richard Koszarski for supplying this reference from his collection.

13. At MGM, Corbaley worked with Irving Thalberg and Louis B. Mayer, and as assistant to Samuel Marx. Marx observes of Corbaley's beginnings that her sojourn at the Palmer Institute of Authorship ("rated an estimable school in the dubious business of teaching amateurs how to write for the screen") preceded her labors at Thomas Ince's studio, from which she moved on to MGM (79).

14. See also Palmer's *The Essentials of Photoplay Writing*, which justifies the process by saying, "This is done so all manuscripts will be uniform in appearance and instantly identified as Palmer Plan material. The quality of Palmer stories being thoroughly established, it is obvious that any synopsis submitted through us will receive immediate consideration" (19).

15. It is impossible to determine how many students the school took of those that applied, but it seems unlikely that any significant culling of applications took place. For example, Manker was quoted as "predict[ing] the number of enrollments will be increased three to four thousand before the year [1920] is out" ("Palmerland," 12). In June 1921, LeBerthon reported the circulation of *The Photodramatist* as thirteen thousand (12); to be sure, it was possible to subscribe to the journal for $2.50 without joining the school.

16. This letter-writing campaign anticipated that organized by the Legion of Decency, discussed in chapter 4, which urged audience members to write to their exhibitors and others connected with the film industry asking that certain objectionable films not be played.

17. The terms for *Judgment of the Storm*, for example, were a minimum advance royalty of one thousand dollars and some unspecified further fraction of the profits ("Palmer Enters Field").

18. The "cutting script" has the same kind of authority as that enjoyed by the continuity script and possibly for the same reasons: both are comparatively "technological" manifestations of screenwriting, in which the blueprint for the film is lauded as an avatar of the engineer-author.

19. See *Variety*'s review of *His Forgotten Wife*, which dismisses the last picture as "one of the series of yarns produced by the Palmer Photoplay School as an inducement to show those who take their scenario courses that there is a production chance for the work they turn out after paying to learn how it is done." Sargent argues that the production scheme more than pays for itself: "The company using this idea loses no money; in fact it is reputed to have cleaned up more than a million dollars in the past few years" (26). Given that Palmerplays had been in production for only a year at this point, however, most of those profits must have come from instruction fees.

20. Indeed, Palmer's literary references straddle the divide between the highbrow (Ibsen) and the popular (Dumas), although this distinction is to some extent an artifact of Modernism.

21. While fan magazines were generally glowing about female contributions to the film industry, they also sometimes sought to disavow the very elements that women found most appealing. During Valentino's reign as most beloved star, for example, *Motion Picture Magazine* and *Motion Picture Classic* printed articles and letters suggesting that fans might pay tribute to alternative, more savory, stars (these columns would sometimes meet with fan rebuttal). Similarly, the incessant contests for most popular actress and actor did not necessarily represent vox populi so much as a way of channeling discourse about what the industry most desired.

4. "Sermons in Screens"

1. In theory, violations of the Production Code could be punished by a $25,000 fine, payable by the offending studio. In practice, however, this penalty was almost never invoked.

2. Charles Musser argues that film first entered the purview of some Protestant groups as "an updated magic lantern or stereopticon, forerunners of the modern slide projector" (71). This observation attaches some Protestants' initial experience of film to earlier traditions of church-sponsored culture, which took the place of activities some denominations frowned upon, such as attendance at theaters. Significantly, the illustrated lecture, which film here replaces, could be minutely tailored to the tastes and interests of the local audience (Musser, 83), a feature notably absent from the autonomous narratives of the sound era.

3. Patricia Zimmermann finds that the rhetoric of good fellowship through amateur filmmaking also worked during this period on behalf of another economically marginal institution, the American family (44).

4. Ironically, there is some evidence that any ambitious program of educational film manufacturing would have required block booking to operate. See Earle W. Hammons's comments in *The Story of the Film* about the operations of his "Educational Pictures, Inc.," which distributed and manufactured short films of many kinds, including comedies. In this transcript of a lecture given at the Harvard Business School in 1927, Hammons acknowledges during the question period that his firm must block book in order to control overhead and anticipate demand (164).

5. See, for example, "Harry Levey Company to Furnish Pictures to Churches Everywhere," which reports that "the Christian Herald [parent organization of the Christian Herald Motion Picture Bureau] is now militantly espousing the cause of a free and unlimited supply of church pictures, with no line drawn except such as good taste and clean morals suggest. In this they take issue with Will H. Hays, who has publicly asserted that the schools should stick to educational subjects, the church to religious films, and that amusement films be restricted to the theatres" (660).

6. Garth Jowett notes that the committee included representatives from such organizations as "the Daughters of the American Revolution, the Boy Scouts of America, the General Federation of Women's Clubs, the International Federation of Catholic Alumnae, the Russell Sage Foundation, the American Library Association, the YMCA and the National Recreation Association" ("Moral Responsibility," 11).

Richard deCordova also mentions as members the American Legion and the American Civic Association ("Ethnography," 99), while Ruth Vasey adds the International Federation of Catholic Alumni, the National Catholic Welfare Congress, the National Congress of Parents and Teachers, and the National Education Association (33). One group that steadfastly remained outside the Hays Office umbrella was the Women's Christian Temperance Union, which campaigned for federal censorship legislation into the 1930s and rejected the Hays Office plan for motion picture "endorsement" as offering film manufacturers carte blanche in the making of "bad" films while conceding nothing to reformers (Parker, 85). For an extended discussion of the relations between the WCTU and the film industry, see Alison Parker's "Mothering the Movies."

7. It was, of course, possible to take the view that "chain" theaters (those owned by the studios) more efficiently communicated patrons' attitudes toward films to the front office, since independent exhibitors might not be moved to canvass their audiences or report major successes or failures to the manufacturer. J. Homer Platten, writing for *The Banker's Magazine*, argues that the chain theaters could be used by independents as a bellwether for the success or failure of a given film, an approach that completely ignores the want of choice arising from block booking (490).

8. Hays apparently wanted to investigate the popularity of religious pictures through some sort of trial. *Motion Picture News* reports that the MPPDA's "representatives [presumably Wallace] have surveyed the entire field of films having to do with these subjects and from several hundred have selected a small group which are being shown in twelve churches in twelve different towns within a radius of 150 miles of New York over a period of twelve consecutive Sundays. The purpose is to learn by this experiment how the programs affect the congregation, whether attendance is increased, how useful a picture may be in religious service. If the results are as anticipated we will organize adequately the demand and the supply" ("Film in Churches," 4 July 1925, 32).

9. The collapse of the Hays Office's cordial relationship with a number of the most prominent public service organizations happened in stages, with such milestones as the defections of the National Congress of Parents and Teachers (1924), the General Federation of Women's Clubs, and the Girl Scouts (both in 1925) (Vasey, 239 n19). For a sustained treatment of this subject, see Carr.

10. For a sustained treatment of this subject, see Carr.

11. Gregory Black observes that "Catholics comprised only one-fifth of the population, but they were heavily concentrated in cities east of the Mississippi River. Chicago was one-half Catholic, as was Boston. New York, Buffalo, Philadelphia, Pittsburgh, Cleveland, and Detroit had sizable Catholic populations" (*Crusade*, 22).

12. For the text of the pledge, see Inglis (122). Curiously, it turns out that raters sometimes relied upon trade journals and other published sources of reviews in the formation of their opinions, so it is possible that a particular film was banned at the urging of a group who had never seen it (Walsh, 97).

13. For example, an attack on indecent movies mounted before the Legion of Decency campaign by a parish in Massachusetts had real teeth: "The parish priests announced that they would refuse parishioners absolution if they patronized the offending theater" (Walsh, 92).

14. T. O'R. Boyle denies the analogy in his "We Are All Censors," published in *The Commonweal* in 1934 (227), on the basis that works appearing on the Index

were listed because of their doctrinal deviations, a connection that no one was claiming for immoral films. For most non-Catholics, one imagines, this distinction was not obvious.

15. Coughlin's diocese appears to have entered quickly into the program of blacklisting, and his highly supportive bishop, Michael Gallagher, was willing to condemn the film industry. Coughlin's failure to condemn film as roundly as he condemned much else, however, gave some Catholics the impression that he had been "bought off" by Hollywood (Walsh 96, 100, 132–33).

16. Indeed, demand for movies during the Depression was surprisingly strong. Citing Floyd B. Odlum, Jowett notes that "between 1929 and 1933 consumer expenditure for shoes fell by 41 percent, and that for food for home consumption by 40 percent, while motion picture theater receipts dropped only 33 percent" (*Film*, 286 n5).

17. Cardinal Mundelein was also a significant force behind the earliest stages of the creation of the Production Code in 1930 because municipal censorship in Chicago was still quite active in the late 1920s and representatives of the Hays Office hoped to enlist Mundelein's support for alternatives to direct censorship (Walsh, 57–59).

18. Walsh notes that Quigley and Breen, who seem not to have cared for Father Lord personally, were further outraged at his beginning to blacklist films in *The Queen's Work* (98). One of the virtues of Walsh's excellent account of the Legion of Decency is the scrupulousness with which he traces the many divisions within Catholic approaches to controlling film content, which tended to belie the impression that Catholics gave to outsiders—namely, of being remarkably unified in their approach.

19. Black details the wealth of publications used by the Catholic Church to reach its members. In addition to a number of national publications such as *The Commonweal* (on the liberal side of the spectrum politically), he lists "*Catholic World, America, Sign,* and *Thought;* the *Ecclesiastical Review,* a journal directed at priests; and *Catholic Digest,* for those who wanted their theology in condensed form" (*Crusade,* 22). Furthermore, he notes, nearly all of the 103 dioceses had their own newsletters (*Crusade,* 22). For evidence of the harnessing of Catholic educational institutions in letter-writing campaigns, see Walsh (99).

20. For a discussion of the ways motion pictures were thought to lead to "sex stimulation," see Herbert Blumer and Philip M. Hauser's contribution to the Payne Fund studies, *Movies, Delinquency, and Crime,* which states that "one of the chief ways by which motion pictures seem, unwittingly, to encourage delinquency in the case of some girls and young women is in arousing sex passion. In the case of many of the sexually delinquent girls studied, the autobiographical evidence shows in an impressive way how passionate love pictures may act as a sexual excitant" (81).

5. Learning to Understand the Foe

1. Smoodin's insightful article focuses primarily upon what the reception of Frank Capra films in high schools may indicate about "some of the tensions in the era's film audience, or at least the adolescent one," especially regarding issues of race relations and gender.

2. Arguably, reformers' emphasis on socially responsible filmmaking above profits modeled a similar ethos for the industry.

3. For a discussion of Dewey's connection to the work of previous educators, see Baker. Dewey's own discussion of Froebel in *The School and Society* suggests that as early as 1897 he was already preoccupied by the association between discrete educational techniques and the need to train children to take up their place as citizens of a democracy (116).

4. According to Garth Jowett, Dewey expressed concern over the possibility that the motion picture would come to replace "occupation" in the curriculum when it was used to teach academic subjects (*Film*, 92). As Dewey also observed, "a closely associated danger is that [motion pictures'] use will, for a time at least, strengthen the idea, already much too strong, that the end of instruction is the giving of information and the end of learning its absorption" (quoted in Jowett, *Film*, 92). Film appreciation and character education both emphasized occupation to the extent that they generated activities designed to give students experience in problem-solving. Where film appreciation may have acquiesced in the kind of passive acquisition of information that worried Dewey was in the more "efficient" or palatable communication of literary or historical content. Yet even here, the film appreciation curriculum emphasized activities.

5. The Payne Fund Studies, commissioned by the Motion Picture Research Council to assess the impact that film attendance might be having upon the young, are also the point at which Charters's antipathy to the film industry is most evident. Jacobs cites Charters and the Rev. William H. Short as the moving spirits behind the publication of Henry James Forman's *Our Movie-Made Children* (1933), written to herald the forthcoming Studies (32). Forman's book was markedly anti-Hollywood, and did not accurately reflect the findings or the attitudes of a number of the Payne Fund researchers. For more information on this last point, see Jowett, Ian Jarvie, and Kathryn Fuller's *Children and the Movies: Media Influence and the Payne Fund Controversy*, which argues that Forman's highly biased presentation of the studies doomed them to premature oblivion.

6. The most significant series is certainly *Photoplay Studies*, on whose editorial board Lewin served. *Photoplay Studies* and the *Group Discussion Guide* were published somewhat irregularly, but the ideal was monthly publication designed to co-ordinate with the national distribution of suitable films. The guides sold for four to ten cents (priced on a sliding scale according to the number bought) and indicate a shrewd conception of the large (and captive) market of students that film appreciation programs might tap. Richard Maltby suggests that guides from the NCTE appeared in "print runs of several hundred thousand" ("Code," 63).

7. There was plenty of cross-fertilization between the two wings. For example, Helen Rand and Richard Lewis's *Film and School: A Handbook in Moving-Picture Evaluation* (1937) appeared under the imprimatur of the NCTE but received the "advice and counsel of Edgar Dale and Sarah McLean Mullen." Dale's film appreciation text was the best-selling volume of the eight published Payne Fund studies (Jowett et al., 101).

8. This conflation of work and play has already appeared, of course, in the discussions of Ruth Fielding and of the Palmer Photoplay materials.

9. Lewin was probably the least hostile to the film industry as a whole, which

appears to consort with his attitude toward vocational training and film as an inherently superior mode of instruction for most topics.

10. Lewin, for one, longed for the day when school *would* compete with the picture palace: "To wire school auditoriums for talking-picture projection, of course, will require the expenditure of many millions of dollars. Hence, such a consummation, though devoutly to be wished, is hardly to be expected in the immediate future, for among nearly 100,000 schools in the United States only a few hundred are as yet equipped to show sound films" (*Photoplay Appreciation*, 3).

11. Elizabeth Pollard is even more explicit about the issue in her *Motion-Picture Study Groups*, where she reports that "one of the most important problems in connection with wholesome movies is the problem of eliminating the evil effects of the compulsory block-booking system.... Organizations such as the Allied States Association of Motion Picture Exhibitors are now working toward the abolition of the compulsory block-booking system. Their materials will afford members of the group interesting talking points" (29).

12. Bulletin boards loomed large in the minds of film appreciators. Mullen states confidently that "almost any English teacher will give her students credit if they work in the maintaining of an attractive bulletin board. At Weequahic High School, the Photoplay Club has the use of sixty square feet of cork bulletin board. Here the students display stills from the best current movies, tear-sheets of the movie page of the Sunday *New York Times*, clippings from other papers and magazines, original drawings made by pupils on film subjects, news items about club activities, lists of other school clubs whose activities are being filmed by the camera committee, theater announcements telling about exceptionally artistic pictures, lists of pictures which are suitable for the whole family," and more (55–56).

13. The periodical that has furnished these examples, *Photoplay Studies*, was a discussion guide for teachers promoting film appreciation; it appeared approximately a month before the release of worthy films.

14. See, for example, *The Motion Picture and the Family* 15 February 1936 on *The Good Earth*, 15 December 1936 on *The Plainsman*, 15 January 1937 on *Maid of Salem*, 15 April 1937 on *The Prince and the Pauper*, and 15 May 1937 on *The Toast of New York*. Each of these issues praises the film in question for good research and compelling production values. But historical or literary value aside, on some level these texts present each film as spectacle.

15. Dale argues in a somewhat more nuanced way that "in almost every portrayal of criminals, they appear ready-made.... Only rarely in the 40 pictures under discussion was there any indication that criminal patterns of behavior develop as a product of a long process of interaction between the individual and the successive social situations in which he lives. Since the pictures fail to portray this continuity of experience which produces the criminal, they cannot justifiably claim that they present an intelligent portrayal of the cause and cure of the crime problem" (*Content*, 130).

16. It is not clear, under these circumstances, what the reader should make of the often repeated information about the cost of components of motion pictures in study guides and other materials bearing Lewin's imprimatur. Sometimes Lewin seems impressed by the expense of film production as an institutional factor in what happens to narratives; sometimes, he merely seems impressed by costs.

17. They might also fail to interest audiences altogether. In his digest of Mortimer Adler's rebuttal of the Payne Fund studies, Raymond Moley delights in noting that Dale "suggests it would be better if the movies were to tell stories about 'the problems of land tenancy, the changing mores of agricultural communities, the fading of pioneer psychology and so on.'" Moley adds, "Imagine looking forward to an evening's entertainment devised by Dr. Dale!" (*Are We Movie Made?*, 29).

18. *Scholastic Magazine*, which used a scale rather similar to that promulgated by the NCTE, asked its young subscribers to send in preprinted "score cards," which they could complete individually or as members of a photoplay study club. The score cards were periodically tabulated, and the results then appeared in *Scholastic* (Mullen, 30). Each *Group Discussion Guide* also contained a rating scale identical to those appearing in Lewin's *Photoplay Appreciation*, presumably for use with the picture under discussion.

19. Janet Staiger discusses these two schools of criticism in a somewhat different context in *Bad Women*, where the issue is how the film narrative motivates the discussion of objectionable behavior. The whole-picture school held that, if properly motivated and ultimately criticized or punished, misbehavior is a legitimate subject; the National Board of Review often took this stance (78–79, 104). As Staiger notes, however, the "parts" or "pointillist" school had its adherents (79). The emphasis in Mones's discussion appears to be more aesthetic than moral, and coincides therefore with the attitudes (if not the approach) of the National Board of Review, characteristically ready to bless a socially unconventional picture of high aesthetic worth.

20. Patricia R. Zimmermann's *Reel Families: A Social History of Amateur Film* is a helpful overview of discourse on this subject. Dale edited a section of *Motion Pictures in Education* on "Film Production in Schools," summarizing a number of articles appearing in educational periodicals about the role of filmmaking by students in a variety of subject areas, from Latin to English and social studies. *Educational Screen* published a monthly column providing technical information called "Film Production in the Educational Field," and *Amateur Movie Makers* (later *Movie Makers*) similarly often carried a column on visual education called "Educational Films," in addition to publishing accounts of successful amateur productions made in schools.

21. See, for example, "Cinema Democracy" by Mina Brownstein, about the Little Screen Players of Boston. While Brownstein praises the communal spirit of the Players and argues that their efforts are democratic, she also notes that the seal of their success is found in the translation of a few of their most noted actors into the ranks of the professionals (42). Significantly, she closes with the suggestion that the possibility of social rise is exactly what hampers the amateur: "When the proficiency and talent trained by amateurs is no longer snapped up by professionals, but conserved by the amateurs themselves, the Cinema Muse will be slated for a new Sunday dress" (43).

22. *The Motion Picture and the Family* provides these examples of the kinds of moral problems derived from films in general release: "*Young America* cites a case of juvenile delinquency in which two boys steal a bottle of medicine from the drug store when the grandmother of one of them is taken seriously ill. In *Alias the Doctor* one young man assumes responsibility for the crime of another, which was committed while the latter was under the influence of liquor" ("Three Reels," 6). The

titles otherwise included *Huckleberry Finn, The Sign of the Cross, Cradle Song, Skippy, Sooky, Tom Sawyer, Tom Brown of Culver, Lucky Dog, There's Always Tomorrow, Wednesday's Child, Gentlemen Are Born,* and *No Greater Glory* (Moley, *Hays Office,* 154).

23. This apparent distrust of stimulation was evidently a general concern to Dewey's student Charles Horton Cooley. Daniel Czitrom reports that "the enlargement of social environment through the media produced 'a more rapid and multitudinous flow of personal images, sentiments, and impulses.' For many persons this brought 'an over-excitation which weakens or breaks down character'" (100). I am grateful to Hilary Radner for bringing this reference to my attention.

24. Joas notes that Dewey objected to a "psychology that believes it has found its object in the establishment of law-like causal relations between environmental stimuli and the organism's reactions. Dewey denies that we can legitimately conceive of actions as additively composed of phases of external stimulation, internal processing of the stimulus, and external reaction. To this 'reflex arc model' he opposes the totality of the action: it is the action that determines which stimuli are relevant within the context defined by the action" (21).

25. For three studies that discover that many adolescents already possess appropriate viewing standards, see Clarence Perry's *The Attitude of High School Students Toward Motion Pictures* (1923), Mary Allen Abbott's "A Study of the Motion Picture Preferences of the Horace Mann High Schools" (1927), and Reba Elizabeth Eaton's unpublished "Motion Picture Preferences of Passaic High School" (1929).

Bibliography

Abbott, Mary Allen. "Motion Picture Classics." *English Journal* (College Ed.) 21.8 (October 1932): 624–28.

———. "A Study of the Motion Picture Preferences of the Horace Mann High Schools." *Teachers College Record* 28.8 (April 1927): 819–35.

Adorno, Theodor W., and Max Horkheimer. *Dialectic of Enlightenment.* 1944. Trans. John Cumming. New York: Herder and Herder, 1972.

"Aimee McPherson Plans to Produce Talk-Sound Films." *Motion Picture News* (12 October 1929): 32A.

Ali, Lorraine. "The Glorious Rise of Christian Pop." *Newsweek* 16 July 2001: 40–44.

"An All-Woman Film Company." *Story World and Photodramatist* (July 1923): 79.

Andrews, George Reid. "The Church and the Motion Picture." *National Board of Review Magazine* (March/April 1926): 9–10+.

"Announce a New Picture Series." *Motion Picture and the Family* 1.1 (15 October 1934): 1+.

Appleton, Victor. *The Motion Picture Chums at Seaside Park, or The Rival Photo Theatres of the Boardwalk.* New York: Grosset & Dunlap, 1913.

———. *The Motion Picture Chums' First Venture, or Opening a Photo Playhouse in Fairlands.* New York: Grosset & Dunlap, 1913.

———. *The Motion Picture Chums' New Idea, or The First Educational Photo Playhouse.* New York: Grosset & Dunlap, 1914.

———. *The Motion Picture Chums' Outdoor Exhibition, or The Film That Solved a Mystery.* New York: Grosset & Dunlap, 1914.

———. *The Motion Picture Chums' War Spectacle, or The Film That Won the Prize.* New York: Grosset & Dunlap, 1916.

———. *The Moving Picture Boys and the Flood, or Perilous Days on the Mississippi.* New York: Grosset & Dunlap, 1914.

———. *The Moving Picture Boys at Panama, or Stirring Adventures along the Great Canal.* New York: Grosset & Dunlap, 1915.

———. *The Moving Picture Boys in Earthquake Land, or Working amid Many Perils.* New York: Grosset & Dunlap, 1913.

———. *The Moving Picture Boys on the War Front, or The Hunt for the Stolen Army Films*. New York: Grosset & Dunlap, 1918.

———. *The Moving Picture Boys, or The Perils of a Great City Depicted*. New York: Grosset & Dunlap, 1913.

Azlant, Edward. *The Theory, History, and Practice of Screenwriting, 1897–1920*. Unpublished Ph.D. diss. Madison: University of Wisconsin, 1980.

Bader, David. Letter appearing in the department "Between Ourselves." *Photodramatist* (April 1921): 13.

Badger, Clarence G. *The Point of Attack, or How to Start the Photoplay*. Los Angeles: Palmer Photoplay Corporation, 1920.

Baker, Melvin C. *Foundations of John Dewey's Educational Theory*. New York: King's Crown (Columbia University), 1955.

Balio, Tino. "Surviving the Great Depression." *Grand Design: Hollywood as a Modern Business Enterprise*. Ed. Tino Balio. Berkeley: University of California Press, 1993. 13–36.

Banta, Martha. *Taylored Lives: Narrative Productions in the Age of Taylor, Veblen, and Ford*. Chicago: University of Chicago Press, 1993.

Barnes, Walter. *The Photoplay as Literary Art*. Newark: Educational and Recreational Guides, 1936.

Barton, Bruce. *The Man Nobody Knows: A Discovery of the Real Jesus*. Indianapolis: Bobbs-Merrill, 1925.

Baum, L. Frank, as "Edith Van Dyne." *Aunt Jane's Nieces Out West*. Chicago: Reilly & Britton, 1914.

Beban, George. *Photoplay Characterization*. Los Angeles: Palmer Photoplay Corporation, 1920.

Billman, Carol. *The Secret of the Stratemeyer Syndicate: Nancy Drew, the Hardy Boys, and the Million Dollar Fiction Factory*. New York: Ungar, 1986.

Black, Gregory D. *The Catholic Crusade Against the Movies, 1940–1975*. New York: Cambridge University Press, 1997.

———. *Hollywood Censored: Morality Codes, Catholics, and the Movies*. New York: Cambridge University Press, 1994.

Blumer, Herbert, and Philip M. Hauser. *Movies, Delinquency, and Crime*. New York: Macmillan, 1933.

"Bold Racketeering of Better Films Bodies Alleged." *Motion Picture News* (12 April 1930): 22+.

Boorstin, Daniel J. *The Image: A Guide to Pseudo-Events in America*. New York: Atheneum, 1971.

Bordwell, David, Janet Staiger, and Kristin Thompson. *The Classical Hollywood Cinema: Film Style and Mode of Production to 1960*. New York: Columbia University Press, 1985.

Bourke, Philippa. "Apocalypse Not: A New Breed of Christian Filmmakers Chooses Heart over Hellfire." *Indiewire.com*, 17 July 2001. 14 December 2001. http://www.indiewire.com/film/biz/biz_010717_ChristianFilm.html.

Boyle, T. O'R. "We Are All Censors." *Commonweal* 21.8 (21 December 1934): 226–28.

Brinkley, Alan. *Voices of Protest: Huey Long, Father Coughlin, and the Great Depression*. New York: Vintage, 1983.

Brownstein, Mina. "Cinema Democracy." *Amateur Movie Makers* 2.7 (July 1927): 20+.

Budd, Mike. "The National Board of Review and the Early Art Cinema in New York: *The Cabinet of Dr. Caligari* as Affirmative Culture." *Cinema Journal* 26.1 (Fall 1986): 3–18.

Bush, Gregory W. *Lord of Attention: Gerald Stanley Lee and the Crowd Metaphor in Industrializing America.* Amherst: University of Massachusetts Press, 1991.

Butsch, Richard. "Introduction: Leisure and Hegemony in America." *For Fun and Profit: The Transformation of Leisure into Consumption.* Ed. Richard Butsch. Philadelphia: Temple University Press, 1990. 3–27.

Cantwell, John J. "Priests and the Motion Picture Industry." *Ecclesiastical Review* 90 (February 1934): 136–46.

Carr, Steven. "The Hollywood Question: America and the Belief in Jewish Control over Motion Pictures before 1941." Unpublished Ph.D. diss. Austin: University of Texas at Austin, 1994.

Christian Filmmaker Ministries. "Our Mission." 14 December 2001. http://www.christianfilmmaker.com/mission.htm.

"Christian Film 'Omega Code' to Be Touted in Churches." Inside Entertainment. 29 August 2001. 16 December 2001. http://alliance.zap2it.com/zap2it/code/movies_toolkit/cda/zp_mt_main_template/1,1009,22-108---3542--,00.html.

"Church-Exhibited Pictures Stir Up Picture Industry." *Variety* 3 January 1924: 21.

"Civic Leaders Deny Receipt of Hays Money." *Motion Picture News* (12 April 1930): 22+.

"'Clean and Wholesome' or 'Evil and Cheap'?" *Motion Picture Monthly* 7.6 (August/September 1931): 6.

Clift, Denison. *Dramatic Suspense in the Photoplay.* Los Angeles: Palmer Photoplay Corporation, 1920.

Coffin, Helen Lockwood. "A Minister as a Movie Maker." *National Board of Review Magazine* (June 1931): 8–9+.

Cohen, Lizabeth. *Making a New Deal: Industrial Workers in Chicago, 1919–1939.* New York: Cambridge University Press, 1990.

Conn, Peter. *The Divided Mind: Ideology and Imagination in America, 1898–1917.* Cambridge: Cambridge University Press, 1983.

Corbaley, Kate. *Selling Manuscripts in the Photoplay Market.* Los Angeles: Palmer Photoplay Corporation, 1920.

Couvares, Francis G. "Hollywood, Main Street, and the Church: Trying to Censor the Movies before the Production Code." *Movie Censorship and American Culture.* Ed. Francis G. Couvares. Washington: Smithsonian, 1996. 129–58.

Cvetkovich, Ann. *Mixed Feelings: Feminism, Mass Culture, and Victorian Sensationalism.* New Brunswick: Rutgers University Press, 1992.

Czitrom, Daniel J. *Media and the American Mind: From Morse to McLuhan.* Chapel Hill: University of North Carolina Press, 1982.

Dale, Edgar. *The Content of Motion Pictures.* New York: Macmillan, 1935.

———. "Film Production in Schools." *Motion Pictures in Education: A Summary of the Literature, Source Book for Teachers and Administrators.* Comp. Edgar Dale, Fannie W. Dunn, Charles F. Hoban, Jr., and Etta Schneider. New York: H. W. Wilson Co., 1938. 271–303.

———. *How to Appreciate Motion Pictures: A Manual of Motion-Picture Criticism Prepared for High-School Students.* New York: Macmillan, 1933.

———. "Teaching Motion Picture Appreciation." *English Journal* (College Ed.) 25.2 (February 1936): 113–20.

"Debunking the Gangster: Deadly Weapon of Ridicule Used by the Screen in Pointing the Moral of Current Crime Pictures Strips Gangdom of Glory, An Interview with Judge Cornelius F. Collins." *The Motion Picture Monthly* 7.1 (January 1931): 2–3.

deCordova, Richard. "Ethnography and Exhibition: The Child Audience, the Hays Office and Saturday Matinees." *Camera Obscura* no. 23 (May 1990): 90–107.

———. *Picture Personalities: The Emergence of the Star System in America.* Urbana: University of Illinois Press, 1990.

Denton, Frances. "Real Folks." *Photoplay* (April 1918): 83–86+.

Dewey, John. *Art as Experience.* New York: Minton, Balch, 1934.

———. *The School and Society.* Chicago: University of Chicago Press, 1915.

Doty, Douglas Z. "The Unpublished Author: A Plea for the Cultural Value of Learning to Write." *Photodramatist* (November 1922): 19–20.

Douglas, Ann. *The Feminization of American Culture.* 1977. New York: Doubleday, 1988.

Doyle, Thomas L. *A Guide to the Discussion and Appreciation of* Men with Wings. *Group Discussion Guide* 3.9 (1938).

Dugan, John E. *A Guide to the Discussion and Appreciation of* The Citadel. *Group Discussion Guide* 3.8 (1938).

Dyer, Richard. *Stars.* London: BFI, 1979.

Eaton, Reba Elizabeth. "Motion Picture Preferences of Passaic High School." 1929. Unpublished mss. Will Hays Papers, Jan. 15–31, 1930 folder. Film M720 pt. 2 reel 3.

Eldridge, Donald A. "Motion Picture Appreciation in the New Haven Schools." *Journal of Educational Sociology* 2.3 (November 1936): 175–83.

Elijah, Frances White. "'How It Happened': Successful Writer of Photoplays Tells How She Did It." *Photoplaywright* (April 1921): 3+.

Emerson, Alice B. *Ruth Fielding and Baby June.* New York: Cupples & Leon, 1931.

———. *Ruth Fielding and Her Great Scenario, or Striving for the Motion Picture Prize.* New York: Cupples & Leon, 1927.

———. *Ruth Fielding and Her Greatest Triumph, or Saving Her Company from Disaster.* New York: Cupples & Leon, 1933.

———. *Ruth Fielding Clearing Her Name, or The Rivals of Hollywood.* New York: Cupples & Leon, 1929.

———. *Ruth Fielding in Talking Pictures, or The Prisoners of the Tower.* New York: Cupples & Leon, 1930.

———. *Ruth Fielding in the Far North, or The Lost Motion Picture Company.* New York: Cupples & Leon, 1924.

———. *Ruth Fielding in the Great Northwest.* New York: Cupples & Leon, 1921.

———. *Ruth Fielding of the Red Mill, or Jasper Parloe's Secret.* New York: Cupples & Leon, 1913.

———. *Ruth Fielding Treasure Hunting, or A Moving Picture That Became Real.* New York: Cupples & Leon, 1923.

Ewen, Stuart. *Captains of Consciousness: Advertising and the Social Roots of the Consumer Culture.* New York: McGraw-Hill, 1976.

Facey, Paul W. *The Legion of Decency: A Sociological Analysis of the Emergence and Development of a Social Pressure Group.* New York: Arno, 1974.

"Facts About Scenario Writing." *Motion Picture* 1.3 (November 1925): 2–4.

"Film in Churches." *Motion Picture News* (4 July 1925): 32.

Fine, Richard. *West of Eden: Writers in Hollywood, 1928–1940*. Washington: Smithsonian, 1993.

Fisher, Robert. "Film Censorship and Progressive Reform: The National Board of Censorship of Motion Pictures, 1909–1922." *Journal of Popular Film* 4.2 (1975): 143–56.

"$500,000 Sought by Woman in Alleged Film Piracy." *Variety* (1 June 1926): 4.

"Florida Woman Wins Scenario Contest Conducted by Goldwyn and Newspaper." *Moving Picture World* (15 April 1922): 721.

Forman, Henry James. *Our Movie-Made Children*. New York: Macmillan, 1933.

Fowler, H. E. *A Guide to the Appreciation of the Historical Photoplay* Victoria the Great. *Photoplay Studies* 3.8 (Series of 1937).

Frutchey, F. P., and Edgar Dale. "Testing Some Objectives of Motion Picture Appreciation." *Educational Research Bulletin* 14.2 (13 February 1935): 34–37.

Fuller, Kathryn H. *At the Picture Show: Small-Town Audiences and the Creation of Movie-Fan Culture*. Washington: Smithsonian, 1996.

Gomery, J. Douglas. "Hollywood, the National Recovery Administration, and the Question of Monopoly Power." *Journal of the University Film Association* 31.2 (Spring 1979): 47–52.

Gossman, Lionel. "History and Literature: Reproduction or Signification." *The Writing of History: Literary Form and Historical Understanding*. Ed. Robert H. Canary and Henry Kozicki. Madison: University of Wisconsin Press, 1978. 3–39.

Gould, Stephen Jay. *The Mismeasure of Man*. New York: Norton, 1981.

Graebner, William. *The Engineering of Consent: Democracy and Authority in Twentieth-Century America*. Madison: University of Wisconsin Press, 1987.

Haber, Samuel. *Efficiency and Uplift: Scientific Management in the Progressive Era, 1890–1920*. Chicago: University of Chicago Press, 1964.

Halsey, William M. *The Survival of American Innocence: Catholicism in the Era of Disillusionment, 1920–1940*. Notre Dame: University of Notre Dame Press, 1980.

Hamburger, Louise. "The Greater Movie." *Drama* 14.4 (January 1924): 138–39.

Hammond, Sue Annis, and Joe Hall. "What Is Appreciative Inquiry?" *Inner Edge Newsletter*. Thin Book Publishing Co. 14 December 2001. http://www.thinbook.com/chap11fromle.html.

Hammons, Earle W. "Short Reels and Educational Subjects." *The Story of the Films*. Ed. Joseph P. Kennedy. Chicago: A. W. Shaw Co., 1927.

Hanmer, Lee F. "Motion Pictures." *Social Work Year Book 1929*. Ed. Fred S. Hall and Mabel B. Ellis. New York: Russell Sage Foundation, 1930. 280–82.

Hansen, Miriam. *Babel and Babylon: Spectatorship in American Silent Film*. Cambridge: Harvard University Press, 1991.

Hark, Ina Rae. "The 'Theater Man' and 'The Girl in the Box Office.'" *Film History* 6 (1994): 178–87.

Harris, Neil. *Cultural Excursions: Marketing Appetites and Cultural Tastes in Modern America*. Chicago: University of Chicago Press, 1990.

"Harry Levey Company to Furnish Pictures to Churches Everywhere." *Moving Picture World* (26 August 1922): 660.

Hawley, Ellis. *The Great War and the Search for a Modern Order: A History of the American People and Their Institutions, 1917–1933*. New York: St. Martin's, 1979.

———. "Three Facets of Hooverian Associationalism: Lumber, Aviation, and Movies, 1921–1930." *Regulation in Perspective: Historical Essays.* Ed. Thomas K. McCraw. Cambridge: Harvard University Press, 1981. 95–123.

Hays Papers. Indiana State Library, Indianapolis.

Hays, Will H. *The Memoirs of Will H. Hays.* Garden City: Doubleday, 1955.

"Hays After Film Fakers." *Variety* (20 December 1923): 1.

Heald, Morrell. "Business Thought in the Twenties: Social Responsibility." *American Quarterly* 13 (Summer 1961): 126–39.

Higashi, Sumiko. *Cecil B. DeMille and American Culture: The Silent Era.* Berkeley: University of California Press, 1995.

"His Forgotten Wife." Film review of 25 June 1924. *Variety Film Reviews,* Volume 2. New York: Garland, 1983.

Hofstadter, Richard. *The Age of Reform: From Bryan to F.D.R.* New York: Vintage, 1955.

Holmes, Joseph L. "Do War Films Inculcate War, or the Will to Peace? Recent Psychological Studies of Such Films on Children Develop Interesting Conclusions." *The Motion Picture Monthly* 6.11 (November 1930): 6, 11.

Hope, Laura Lee. *The Moving Picture Girls at Oak Farm, or Queer Happenings While Taking Rural Plays.* New York: Grosset & Dunlap, 1914.

———. *The Moving Picture Girls at Rocky Ranch, or Great Days among the Cowboys.* Cleveland: World, 1914.

———. *The Moving Picture Girls in War Plays, or The Sham Battles at Oak Farm.* New York: Grosset & Dunlap, 1916.

———. *The Moving Picture Girls, or First Appearances in Photo Dramas.* New York: Grosset & Dunlap, 1914.

———. *The Moving Picture Girls Snowbound, or The Proof on the Film.* New York: Grosset & Dunlap, 1914.

———. *The Moving Picture Girls under the Palms, or Lost in the Wilds of Florida.* New York: Grosset & Dunlap, 1914.

Horne, Herman Harrell. *The Democratic Philosophy of Education: Companion to Dewey's Democracy and Education.* New York: Macmillan, 1932.

Hustwick, Alfred. "Motion Pictures—To-day and To-morrow: 5—A Plan to Revive the Motion Picture Art." *Film Spectator* (11 December 1926): 13.

Huyssen, Andreas. "Mass Culture as Woman: Modernism's Other." *Studies in Entertainment: Critical Approaches to Mass Culture.* Ed. Tania Modleski. Bloomington: Indiana University Press, 1986. 188–207.

Inglis, Ruth A. *Freedom of the Movies: A Report on Self-Regulation from The Commission on Freedom of the Press.* Chicago: University of Chicago Press, 1947.

"'The Inner Sight' Is Sold to Ince for $500." *Photoplaywright* (October 1920): 7.

Inness, Sherrie A. "The Feminine En-gendering of Film Consumption and Film Technology in Popular Girls' Serial Novels, 1914–1931." *Journal of Popular Culture* 29.3 (Winter 1995): 169–82.

Jackson, Phyllis. "One Hundred Movie Critics." *Junior Red Cross Journal* (September 1933): 8–10.

Jacobs, Lea. "Reformers and Spectators: The Film Education Movement in the Thirties." *Camera Obscura* no. 22 (January 1990): 29–49.

———. *The Wages of Sin: Censorship and the Fallen Woman Film, 1928–1942.* Madison: University of Wisconsin Press, 1991.

Joas, Hans. *Pragmatism and Social Theory.* Chicago: University of Chicago Press, 1993.

Johnson, Deidre. *Edward Stratemeyer and the Stratemeyer Syndicate*. New York: Twayne, 1993.

———. *Stratemeyer Pseudonyms and Series Books: An Annotated Checklist of Stratemeyer and Stratemeyer Syndicate Publications*. Westport: Greenwood, 1982.

Johnson, Sheldon Krag. "Breaking in from the Top: A Review of the $10,000 Prize Photoplay 'Broken Chains.'" *Photodramatist* (January 1923): 21–24+.

Jowett, Garth. *Film: The Democratic Art*. Boston: Little, Brown, 1976.

———. "Moral Responsibility and Commercial Entertainment: Social Control in the United States Film Industry, 1907–1968." *Historical Journal of Film, Radio, and Television* 10.1 (1990): 3–31.

———, Ian C. Jarvie, and Kathryn H. Fuller. *Children and the Movies: Media Influence and the Payne Fund Controversy*. Cambridge: Cambridge University Press, 1996.

"Judgment of the Storm." Film review of 31 January 1924. *Variety Film Reviews*, Volume 2. New York: Garland, 1983.

Katterjohn, Monte. *How to Write and Market Moving Picture Plays*. Boonville, IN: Photoplay Enterprise Association, n.d.

Kelly, Mary. "Do Dramas Cater Exclusively to Women?" *Moving Picture World* (11 February 1922): 615.

Keyser, Les, and Barbara Keyser. *Hollywood and the Catholic Church: The Image of Roman Catholicism in American Movies*. Chicago: Loyola University Press, 1984.

Klein, Marcus. *Easterns, Westerns, and Private Eyes: American Matters, 1870–1900*. Madison: University of Wisconsin Press, 1994.

Koon, Cline M. *Motion Pictures in Education in the United States: A Report Compiled for the International Congress of Educational and Instructional Cinematography*. Chicago: University of Chicago Press, 1934.

Koszarski, Richard. *An Evening's Entertainment: The Age of the Silent Feature Picture, 1915–1928*. History of the American Cinema vol. 3. Ed. Charles Harpole. New York: Scribner's, 1990.

Krows, Arthur Edwin. "Literature and the Motion Picture." *The Annals of the American Academy* 128 (November 1926): 70–73.

Kummer, Frederic Arnold. "The Rejected One." *Photoplay* (March 1918): 45–49+.

Lane, Tamar. *What's Wrong with the Movies?* Los Angeles: Waverly, 1923.

Law, Frederick Houk. *A Guide to the Study of the Screen Version of* Mutiny on the Bounty. *Photoplay Studies* 1.5 (November 1935).

Leach, Eugene. "Mastering the Crowd: Collective Behavior and Mass Society in American Social Thought, 1917–1939." *American Studies* 27.1 (Spring 1986): 99–114.

Lears, Jackson. *Fables of Abundance: A Cultural History of Advertising in America*. New York: Basic, 1994.

———. "From Salvation to Self-Realization: Advertising and the Therapeutic Roots of the Consumer Culture, 1880–1930." *The Culture of Consumption: Critical Essays in American History, 1880–1980*. Ed. Richard Wrightman Fox and T. J. Jackson Lears. New York: Pantheon, 1983. 3–38.

———. *No Place of Grace: Antimodernism and the Transformation of American Culture, 1880–1920*. New York: Pantheon, 1981.

LeBerthon, Ted. "This Side of Nirvana." *Photodramatist* (June 1921): 11–12.

———. "This Side of Nirvana." *Photodramatist* (July 1921): 13–14.

LeSourd, Howard M. "The Films in a New Field." *National Board of Review Magazine* 10.3 (March 1935): 4–6.

———. "Plans for Utilizing 'Secrets of Success.'" *Motion Picture and the Family* 2.1 (15 September 1935): 6.

———. "The Use of Motion Pictures in Character Education." *National Board of Review Magazine* 8.1 (January 1933): 4–5.

Levine, Lawrence. *Highbrow/Lowbrow: The Emergence of Cultural Hierarchy in America*. Cambridge: Harvard University Press, 1988.

Lewin, William. "First Experiments with Talkies in American Schools." *Educational Screen* 9.2 (February 1930): 41–43.

———. *Photoplay Appreciation in American High Schools*. New York: D. Appleton-Century Co., 1934.

———. "Photoplays for Vocational Guidance." *Educational Screen* 6.1 (December 1927): 452–54.

———. "Standards of Photoplay Appreciation." *English Journal* (College Ed.) 21.10 (December 1932): 799–810.

———. "Teachers Hail the Talkies." *Educational Screen* 8.10 (December 1929): 295–96+.

———. *Teachers' Key to Accompany the Student's Guide to Critical Appreciation of the Photoplay Version of Stevenson's* Treasure Island." New York: National Council of Teachers of English, 1934.

———. "Where Does Progress Lie in the Development of Educational Talking Pictures? (II)." *Educational Screen* 9.1 (January 1930): 8–9.

Lewis, Sinclair. *Babbitt*. 1922. New York: Signet, 1980.

Lippmann, Walter. *Public Opinion*. 1922. New York: Free, 1965.

"The Little Men Behind the Big Screen." By a Producer of Moving Pictures. *Collier's* (30 September 1922): 11–12+.

Little Stories of Success. Los Angeles: Palmer Photoplay Corporation, 1922.

Lord, Daniel A., S.J. *I Can Read Anything*. St. Louis: Queen's Work, 1930.

Lynd, Robert S., and Helen Merrell Lynd. *Middletown: A Study in Contemporary American Culture*. New York: Harcourt, Brace and Co., 1929.

———. *Middletown in Transition: A Study in Cultural Conflicts*. New York: Harcourt, Brace and Company, 1937.

MacPherson, Jeanie. "Functions of the Continuity Writer." *Opportunities in the Motion Picture Industry and How to Qualify for Positions in Its Many Branches*. Los Angeles: Photoplay Research Society, 1922. 25–26.

———. *The Necessity and Value of Theme in the Photoplay*. Los Angeles: Palmer Photoplay Corporation, 1920.

Maltby, Richard. "'Baby Face' or How Joe Breen Made Barbara Stanwyck Atone for Causing the Wall Street Crash." *Screen* 27.2 (March–April 1986): 27–45.

———. *Harmless Entertainment: Hollywood and the Ideology of Consensus*. Metuchen: Scarecrow, 1983.

———. "*The King of Kings* and the Czar of All the Rushes: The Propriety of the Christ Story." *Screen* 31.2 (Summer 1990): 188–213.

———. "The Production Code and the Hays Office." *Grand Design: Hollywood as a Modern Business Enterprise, 1930–1939*. Ed. Tino Balio. Berkeley: University of California Press, 1993. 37–72.

———. "'To Prevent the Prevalent Type of Book': Censorship and Adaptation in Hollywood, 1924–1934." *Movie Censorship and American Culture*. Ed. Francis G. Couvares. Washington: Smithsonian, 1996. 97–128.

———. "Sticks, Hicks, and Flaps: Classical Hollywood's Generic Conception of Its Audiences." Unpublished paper delivered at the Commonwealth Fund Conference in American History on Hollywood and Its Spectators: The Reception of American Films, 1895–1995, 12–14 February 1998, University College, London.

Manker, Roy. Undated circular letter to subscribers from the Palmer Photoplay Corporation, ca. 1924.

Marchand, Roland. *Advertising the American Dream: Making Way for Modernity, 1920–1940*. Berkeley: University of California Press, 1985.

Marx, Samuel. *Mayer and Thalberg: The Make-Believe Saints*. New York: Random House, 1975.

May, Lary. *Screening Out the Past: The Birth of Mass Culture and the Motion Picture Industry*. Chicago: University of Chicago Press, 1980.

May, Mark A. "Educational Possibilities of Motion Pictures." *Journal of Educational Sociology* 2.3 (November 1936): 149–60.

Mayne, Judith. "Immigrants and Spectators." *Wide Angle* 5.2 (1983): 33–40.

"Mentality Test for Mass. Censors." *Motion Picture News* (28 December 1929): 24.

"Methodist-Made Pictures." *Variety* 74.3 (5 March 1924): 1+.

Michaels, Walter Benn. "An American Tragedy, or the Promise of American Life." *Representations* 25 (Winter 1989): 71–98.

Milliken, Carl E. "Increasing General Usefulness of Films: Educational Pictures, Religious Pictures, Medical Pictures, and Films in Industry Are Given Encouragement." *The Motion Picture* 4.8 (August 1928): 6–7.

"Ministers to Aid Fox Picture: Will Assist in Preparation of Story as Well as Appear in 'Thank U'." *Motion Picture News* 31.5 (31 January 1925): 429.

Moley, Raymond. *Are We Movie Made?* New York: Macy-Masius, 1938.

———. *The Hays Office*. Indianapolis: Bobbs-Merrill, 1945.

Mones, Leon. *A Guide to the Appreciation of Music for Madame*. *Photoplay Studies* 3.6 (Series of 1937).

Moon, Michael. "'The Gentle Boy from the Dangerous Classes': Pederasty, Domesticity, and Capitalism in Horatio Alger." *Representations* 19 (Summer 1987): 87–110.

Moracin, Theodore. "The Transmuters of Dreams." *Photodramatist* (June 1921): 13–15.

Morey, Anne. "Priests in Charge: The Filmic Redemption of the Urban Boy." Paper presented at the Children's Literature Association meeting, Omaha, NE, June 1997.

———. "'Would You Be Ashamed to Let Them See What You Have Written?' The Gendering of Photoplaywrights, 1913–1923." *Tulsa Studies in Women's Literature* 17.1 (Spring 1998): 83–99.

———, and Claudia Nelson. "Mass Reading/Mass Media: Juvenile Series Fiction about the New Technology, 1910–1922." Paper presented at the South Central Modern Language Association meeting, San Antonio, TX, November 1996.

"Motion Picture Study Is No Sinecure in East Orange." *Motion Picture and the Family* 1.9 (15 June 1935): 1.

"Motion Pictures and Art, Science and Religion: A Survey of the Part Motion Pictures Play in the Economic and Cultural Life of the Community." *The Motion Picture* 5.12 (December 1929): 4–5.

"The Movies: A Colossus That Totters." *Bookman* (February 1919): 655–56.

"Much Praise for 'Secrets of Success' Films." *Motion Picture and the Family* 1.4 (15 January 1935): 2.

Mullen, Sarah McLean. *How to Judge Motion Pictures: A Pamphlet for High School Students.* New York: Scholastic, 1934.

Musser, Charles. *The Emergence of Cinema: The American Screen to 1907.* New York: Scribner's, 1990.

──────, and Carol Nelson. *High-Class Moving Pictures: Lyman H. Howe and the Forgotten Era of Traveling Exhibition, 1880–1920.* Princeton: Princeton University Press, 1991.

Nackenoff, Carol. *The Fictional Republic: Horatio Alger and American Political Discourse.* New York: Oxford University Press, 1994.

National Board of Review Papers. Rare Books and Manuscripts Division, New York Public Library, New York, NY.

Nelson, Richard Alan. "Propaganda for God: Pastor Charles Taze Russell and the Multi-Media *Photo-Drama of Creation* (1914)." *An Invention of the Devil? Religion and Early Cinema: Proceedings of the First International Domitor Conference, Québec, June 1990.* Ed. Roland Cosandey and André Gaudreault. Sainte-Foy: Les Presses de l'Université Laval, 1992. 230–55.

"Noted Teacher Praises New Character Films." *Motion Picture and the Family* 1.2 (15 November 1934): 6.

O'Higgins, Harvey. "Study Your Audience." *Photodramatist* (July 1921): 9.

Ohmer, Susan. "Measuring Desire: George Gallup and Audience Research in Hollywood." *Journal of Film and Video* 43.1–2 (Spring–Summer 1991): 3–28.

Orndorff, Marguerite. *A Guide to the Study of the Screen Version of* Captains Courageous. *Photoplay Studies* 3.4 (April 1937).

Palmer, Frederick. *Author's Fiction Manual.* Hollywood: Palmer Institute of Authorship, 1924.

──────. *The Essentials of Photoplay Writing.* Los Angeles: Palmer Photoplay Corporation, 1921.

──────. Letter of 29 August 1922 to F. C. Baker, in the possession of Janet Staiger.

──────. *Palmer Plan Handbook.* Los Angeles: Palmer Photoplay Corporation, 1918, 1921.

──────. "The Perturbation of the Misinformed." *Bookman* (July 1919): 590–91.

──────. *Photoplay Plot Encyclopedia.* Hollywood: Palmer Photoplay Corporation, 1922.

──────. *Self-Criticism.* Hollywood: Palmer Photoplay Corporation, 1923.

──────. "Today and Tomorrow." *Story World and Photodramatist* (March 1923): 76.

──────. "Whence Future Photoplays? The Demands of the Screen Not Met by Fictionists." *Forum* (August 1919): 189.

"Palmerland." *Photoplaywright* (November 1920): 12.

"Palmer Photoplay Corporation Enters Film Producing Field." Circular sent to students. Summer 1922.

"Palmer Photoplay Opens Office in New York City." *Moving Picture World* (11 February 1922): 613.

Parker, Alison M. "Mothering the Movies: Women Reformers and Popular Culture." *Movie Censorship and American Culture.* Ed. Francis G. Couvares. Washington: Smithsonian, 1996. 73–96.

"Pastor Gives Weekly Show." *Motion Picture and the Family* (15 June 1935): 3.

Patterson, Frances Taylor. "Motion Picture Review—How It Should Function to Reflect Public Preference." *National Board of Review Magazine* (June 1934): 8–10.

Pells, Richard H. *Radical Visions and American Dreams: Culture and Social Thought in the Depression Years.* New York: Harper & Row, 1973.

Peltret, Elizabeth. "My Experience in an Academy of Motion Picture Art." *Photoplay* (February 1919): 57–58.

Peters, Charles C. *Motion Pictures and Standards of Morality.* 1933. New York: Arno, 1970.

Peyser, Marc. "God, Mammon, and 'Bibleman.'" *Newsweek* 16 July 2001: 45–48.

Philips, Henry Albert. *The Universal Plot Catalogue.* Springfield, MA: Home Correspondence School, 1916.

Platten, J. Homer. "Motion Pictures—A New Public Utility?" *Banker's Magazine* 113.4 (October 1926): 453–58, 486–92.

Pollard, Elizabeth Watson. *Motion-Picture Study Groups: Handbook for the Discussion Leader.* Columbus: Bureau of Educational Research, Ohio State University (in cooperation with the Payne Fund), 1934.

———. *Teaching Motion-Picture Appreciation: A Manual for Teachers of High-School Classes.* Columbus: Bureau of Educational Research, Ohio State University, 1935.

"Psychologist to Assist Universal Production." *Motion Picture* 5.1 (January 1929): 7.

Pumphrey, Martin. "The Flapper, the Housewife, and the Making of Modernity." *Cultural Studies* 1 (May 1987): 179–94.

Radinoff, Florence. *The Photoplaywright's Handy Text-Book.* New York: Manhattan Motion Picture Institute, 1913.

"Raising the Standards of Demand: Community Leaders Tell of Constructive Work for Better Pictures through Cooperation and Education." *Motion Picture Monthly* 7.2 (February 1931): 10–12.

Rand, Helen Miller. "Motion Pictures: A Social and Educational Force." *Journal of Educational Sociology* 2.3 (November 1936): 164–65.

Rand, Helen, and Richard Lewis. *Film and School: A Handbook in Moving-Picture Evaluation.* New York: D. Appleton-Century Co., 1937.

Rosenbloom, Nancy J. "In Defense of the Moving Pictures: The People's Institute, the National Board of Censorship and the Problem of Leisure in Urban America." *American Studies* 33.2 (Fall 1992): 41–60.

———. "Progressive Reform, Censorship, and the Motion Picture Industry, 1909–1917." *Popular Culture and Political Change in Modern America.* Ed. Ronald Edsforth and Larry Bennett. Albany: State University of New York Press, 1991. 41–59.

Rosenzweig, Roy. *Eight Hours for What We Will: Workers and Leisure in an Industrial City, 1870–1923.* New York: Cambridge University Press, 1983.

Ross, Steven J. *Working-Class Hollywood: Silent Film and the Shaping of Class in America.* Princeton: Princeton University Press, 1998.

Ryan, Alan. *John Dewey and the High Tide of American Liberalism.* New York: Norton, 1995.

Ryan, Don. "Can Scrub Women Get Rich Writing Scenarios? An Unbiased Inquiry into the Activities of the Palmer Photoplay Corporation." *Don Ryan's Magazine* (n.d. [May 1923?]): 7–11.

Sargent, Epes Winthrop. "Flimflamming the Film Fans." *Woman's Home Companion* (November 1924): 26+.

Sarver, Charles. "How to Write a Movie—And How Not to Write One." *Woman's Home Companion* (May 1920): 9.

Schultheiss, John. "The 'Eastern' Writer in Hollywood." *Cinema Journal* 11.1 (Fall 1971): 13–47.

"Send No Money to Get This Book." Palmer Photoplay circular, n.d.

[Shenton, Herbert]. *The Public Relations of the Motion Picture Industry.* New York: Federal Council of the Churches of Christ in America, 1931.

Sheridan, Marion C. "Rescuing Civilization through Motion Pictures." *Journal of Educational Sociology* 2.3 (November 1936): 166–74.

Singer, Ben. "Female Power in the Serial-Queen Melodrama: The Etiology of an Anomaly." *Camera Obscura* 22 (January 1990): 90–129.

Skinner, R. Dana. "When Hollywood Turns Holy Land." *Commonweal* 6.1 (11 May 1927): 14–15.

Smith, Kerry, and Irene Lemon. "Learning through Film-Making." *Teachers College Record* 39.3 (December 1937): 207–17.

Smoodin, Eric. "'The Moral Part of the Story Was Great': Frank Capra and Film Education in the 1930s." *Velvet Light Trap* 42 (Fall 1998): 20–35.

"Social Workers Rate Character Films a Success." *Motion Picture and the Family* 1.9 (15 June 1935): 1+.

Springer, John Parris. "Hollywood Fictions: The Cultural Construction of Hollywood in American Literature, 1916–1939." Unpublished Ph.D. diss. Iowa City: University of Iowa, 1994.

Staiger, Janet. *Bad Women: Regulating Sexuality in Early American Cinema.* Minneapolis: University of Minnesota Press, 1995.

———. "Conclusions and New Beginnings." *An Invention of the Devil? Religion and Early Cinema: Proceedings of the First International Domitor Conference, Québec, June 1990.* Eds. Roland Cosandey, André Gaudreault, Tom Gunning. Sainte-Foy: Les Presses de l'Université Laval, 1992. 353–60.

———. *Interpreting Films: Studies in the Historical Reception of American Cinema.* Princeton: Princeton University Press, 1992.

———. "Mass-Produced Photoplays: Economic and Signifying Practices in the First Years of Hollywood." *Wide Angle* 4.3 (1981): 12–27.

———. "'Tame' Authors and the Corporate Laboratory: Stories, Writers, and Scenarios in Hollywood." *Quarterly Review of Film Studies* 8.4 (Fall 1983): 33–45.

Sterner, Alice P., and W. Paul Bowden. *A Course of Study in Motion-Picture Appreciation.* Newark: Educational and Recreational Guides, 1936.

Stevens, Jeanne. "Mediocre Pictures—the Remedy." *Photodramatist* (September 1922): 17–18.

Stokes, Cecil A. "Films for the Church: A New Role for the Amateur." *Amateur Movie Makers* 2.9 (September 1927): 18–19.

"StoryCraft 3.0 Software." Advertisement. *StoryCraft Writer's Software and Writers Resources.* 27 November 2000. 14 December 2001. http://www.writerspage.com/software.html.

Studlar, Gaylyn. "Discourses of Gender and Ethnicity: The Construction and De(con)struction of Rudolph Valentino as Other." *Film Criticism* 13.2 (Winter 1989): 18–35.

———. "The Perils of Pleasure? Fan Magazine Discourse as Women's Commodified Culture in the 1920s." *Wide Angle* 13.1 (January 1991): 6–33.

Susman, Warren I. *Culture as History: The Transformation of American Society in the Twentieth Century.* New York: Pantheon, 1984.

Theim, Kelly. "All-Stars in Action." *The Amherst Student Online*. 19 September 2001. 14 December 2001. http://www.halogen.note.amherst.edu/~astudent/2001-2002/issue03/features/02.html.

Thrasher, Frederic. "What Can Research Prove?" *National Board of Review Magazine* (April 1934): 8–13.

"Three Reels Added to 'Secrets of Success.'" *Motion Picture and the Family* 1.3 (15 December 1934): 6.

Tichi, Cecelia. *Shifting Gears: Technology, Literature, Culture in Modernist America*. Chapel Hill: University of North Carolina Press, 1987.

Tidden, Fritz. "Keeping in Personal Touch." *Moving Picture World* (10 June 1922): 553.

"2500 Schools Put Photoplay Books in Circulation." *Motion Picture and the Family* 2.2 (15 October 1935): 6.

"Unsolicited Manuscripts Not Read." *Variety* (22 August 1928): 6.

Uricchio, William, and Roberta E. Pearson. *Reframing Culture: The Case of the Vitagraph Quality Films*. Princeton: Princeton University Press, 1993.

Vasey, Ruth. *The World According to Hollywood, 1918–1939*. Madison: University of Wisconsin Press, 1997.

Vaughn, Stephen. "Financiers, Movie Producers and the Church: Economic Origins of the Production Code." *Current Research in Film: Audiences, Economics, and Law*. Volume 4. Ed. Bruce A. Austin. Norwood, NJ: Ablex Publishing Corp., 1988.

Wagner, Rob. "Nailing a Fallacy." *Photoplaywright* (January 1921): 3.

Walsh, Frank. *Sin and Censorship: The Catholic Church and the Motion Picture Industry*. New Haven: Yale University Press, 1996.

Wayne, Hamilton. "Breaking the Shackles of Silence." *Overland Monthly and Out West Magazine* (July 1925): 257–59.

Westbrook, Robert B. *John Dewey and American Democracy*. Ithaca: Cornell University Press, 1991.

"Where the Blame Lies for Movie 'Sex-Stuff.'" *Literary Digest* (12 February 1921): 28–29.

"The White Sin." Film review of 14 May 1924. *Variety Film Reviews*, Volume 2. New York: Garland, 1983.

"Why I Am Ashamed of the Movies." *Collier's* (16 September 1922): 3–4.

Wiebe, Robert H. *The Search for Order, 1877–1920*. New York: Hill and Wang, 1967.

Winter, Alice Ames. "A Little More Wisdom Please." *Motion Picture* 6.3 (March 1930): 3+.

———. "Some Facts about the Previewing Service." *Motion Picture* 5.12 (December 1929): 3+.

Wright, William Lord. "Photoplay Authors Real and Near." *New York Dramatic Mirror* (9 September 1916): 43.

Yost, Edna. "The Fifty-Cent Juveniles." *Publishers Weekly* (18 June 1932): 2405–8.

———. "Who Writes the Fifty-Cent Juveniles?" *Publishers Weekly* (20 May 1933): 1595–98.

"You'll Get What You Ask For." By a Producer of Moving Pictures. *Collier's* (18 November 1922): 9–10+.

Young, Donald. "Social Standards and the Motion Picture." *Annals* (November 1926): 146–50.

Zimmermann, Patricia R. *Reel Families: A Social History of Amateur Film*. Bloomington: Indiana University Press, 1995.

Index

Abbott, Mary Allen, 173
Acting: accident vs., 55–59
Act One Writing, 197
Actors, 13, 94, 191; audience and, 26, 54, 109; filmmaking and, 58; improving, 52, 53, 58
Adams, Henry, 157
Adams, William T., 39
Adaptation, 72, 73, 101, 102, 126, 165, 167, 179
Adorno, Theodor, 40
Adventure: narrative and, 91
Advertising, 66, 71, 72, 76, 82, 96, 125, 144, 158, 188, 194–95, 205n10; self-commodification and, 83–92
Agents: film industry and, 77
AI. *See* Appreciative Inquiry
Alger, Horatio, 202n4; stories by, 39–43, 46; work of, 40–41, 62–63
Alias the Doctor (movie), 211n22
Allied States Association of Motion Picture Exhibitors, 210n11
Amateur Movie Makers, 174, 211n20
America (periodical), 208n19
American Civic Association, 207n6
Americanization, 71, 173
American Telephone and Telegraph (AT&T), 129
Andrews, George Reid, 117, 128, 145, 147; *King of Kings* and, 127; on religion/ social adjustment, 146

Angels with Dirty Faces (movie): fighting priest in, 143
Angelus Productions, 117
Annals (AAPS), 9
Antagonists, 38, 61
Anti-Semitism, 129–30, 135
Appleton, Victor, 36, 203n8. *See also* Moving Picture Boys
Appreciative Inquiry (AI): described, 194
Aquinas, Thomas, 133
Arousal: character education and, 183, 184
Art, 1, 6, 79, 132, 147, 183; authority of, 77–78; contemplation of, 25, 170; film and, 25, 57; literary, 9, 105; textual strategies of, 27
Associationalism, 134, 153
Astor, Frank, 182
Atlas, Charles, 68
Attendance, 6, 10, 189
Audience, 105, 106, 122, 130, 141, 146, 173; actors and, 26, 54; authors and, 96; consumption by, 109; desire of, 54; disciplined, 25; film industry and, 1, 3, 6, 10, 22, 28, 30, 117, 131; Palmer and, 118–19; power of, 61–63; producer and, 71, 114; segmentation of, 11; stupidity of, 8; text and, 27; understanding, 61, 71, 133; validation of, 50–54

Aunt Jane's Nieces Out West (Baum), 36, 41, 44, 45, 55, 56, 57, 61
Authority, 14–17, 77–78, 167, 188; challenges to, 122; cultural, 78, 138; democracy and, 189
Authorship, 100, 103, 111, 190; audience and, 96; alternative form of, 14; Protestants and, 116–20; secrets of, 73
Avant-garde: mass culture and, 108
Azlant, Edward: on Supreme Court ruling, 76

Babbitt (Lewis), 70
Bader, David, 95
Badger, Clarence G., 104
Banker's Magazine, The, 207n7
Banta, Marta: on narratives, 83; on production, 16, 108–9
Barnes, Elmer Tracey. *See* Motion Picture Chums
Barnes, Walter, 177; on photoplay, 160; on plausibility, 168; on social reality, 168
Barton, Bruce, 18, 202n8
Baum, L. Frank. *See Aunt Jane's Nieces Out West*
Benedict, Ruth, 139
Ben-Hur (Wallace), 125, 126, 165
Bias: problems with, 78
Bibleman, 195
Bildungsroman, 193
Billman, Carol, 67, 202n1
Binet intelligence test, 8
Black, Gregory, 207n11, 208n19
Blacklisting, 131, 208n18
Blind selling, 198
Block booking, 122, 165, 198, 207n7, 210n11
Blockbuster: distribution by, 198
Blue laws, 117, 123
Board of Censorship, 29. *See also* National Board of Review
Bookman, The, 78
Boorstin, Daniel: on stars, 68
Bourke, Philippa, 197
Box office, 63, 146, 196; battle of, 31; excesses of, 138; leverage at, 3, 102; mystique of, 15, 111; plebiscite at, 7–8,

15, 53, 97, 108, 127, 141, 190, 198; pressure at, 191, 198; religion and, 135; returns at, 14, 107, 112, 130
Boycotts, 198
Boyle, T. O'R., 140, 207n14
Boys Town (movie): fighting priest in, 143
Breen, Joseph, 121, 136, 143, 189, 208n18; PCA and, 113, 135
Breen Office, 29
Brinkley, Alan: on populist rhetoric, 201n1
Broken Chains (Kimball), 89, 90; approval for, 98–99
Bronx Public School 82, 186
Bureau of Educational Research (OSU), Dewey and, 158

Callahans and the Murphys, The (movie): reception of, 143
Cantwell, John J.: rhetoric of, 128, 129–30, 135
Capitalism, 15, 44–45
Capra, Frank, 208n1
Captains Courageous (movie), 166
Catholic Action, 143
Catholic Church: authority of, 128–45; cultural profile of, 134; film industry and, 114, 130, 144, 145, 154; filmmaking and, 116, 117, 141–42; Hays Office and, 135; Hollywood and, 128, 140–41, 144, 162; Legion of Decency and, 130–31; natural law and, 141, 169; opposition from, 29, 198; PCA and, 28, 113; Production Code and, 11, 169, 189; social activism and, 133, 134, 145; social order and, 157
Catholic Digest, 208n19
Catholic reformers, 28, 113, 114; audience taste and, 133; film industry and, 150; Protestant reformers and, 24
Catholic World, 208n19
Censorship, 17, 22, 28, 29, 31, 140; alternatives to, 208n17; artificial, 164; campaign for, 207n6; conceptualization of, 30, 163; Hollywood and, 15; natural, 164; Protestant opposition to, 29; public morals and, 8; threat of, 112; working-class, 28

Central High School, 165, 173

Character education, 3, 61, 149, 181–82, 193, 209n4; arousal and, 183, 184; curricula for, 28, 151; descent of, 194; film appreciation and, 2, 169, 182, 184, 189; film text and, 151; religion and, 152; social friction and, 181

Character education movement, 110, 150, 178, 178–87; censorship and, 163; culture of, 183–84; Dewey and, 155; pedagogy and, 154

Charters, Werret Wallace, 158–59, 209n5

Chase, Canon, 30

Chesnaye, Mrs., 91

Chicago Daily News, 89

Children: depiction of war and, 149; fiction of, 202n3; film industry and, 66–67, 173; in managerial roles, 47; Modernism and, 132; protecting, 119

Christian Filmmaker Ministries, 196

Christian Herald Motion Picture Bureau, 206n5

Christianity: film industry and, 196, 197, 198

Church and Drama Association, King of Kings and, 127

"Cinema Democracy" (Brownstein), 211n20

Cinematography Study Group, 159

Citadel, The: discussion guide for, 168

Citizenship, 19, 28, 151, 171

Civic affairs, participation in, 24

Civic organizations, 127

Civilization, 177, 180, 184

Clergy: advisory role of, 6; literary marketplace and, 116. See also Religion

Clift, Denison, 104–5

Coffin, Helen Lockwood, 119

Cohen, Sydney S., 124–25

Collaboration, 92–97, 100

Collier, John, 30

Collier's, 9, 10–11, 13

Columbia University: extension program at, 80

Comet Film Company, 47

Committee on Public Relations (MPPDA), 14, 124

Committee on Social Values in Motion Pictures, 179, 180, 182

Commodification, 54, 63, 72, 91, 96, 103, 145, 199; desirability and, 18–19; issue of, 195. See also Self-commodification

Commodities, 27, 36, 54, 84

Commonweal, The, 140, 208n19; on representation of Christ, 142

Communication, 83, 157

Community Motion Picture Bureau, 121

Conn, Peter: on reformers, 153

Consumer culture, 3, 24, 33–34, 116

Consumers, 27, 166, 193; film appreciation movement and, 163; film industry and, 34; response by, 22, 26, 160

Consumption, 20, 47, 110, 152, 165; audience and, 109; commercialized, 22; consequences of, 28; controlling, 22; cultural, 23, 139; democracy and, 190; film industry and, 33–34; idols of, 49, 65; production and, 172; public, 39; women and, 32, 33

Content: concern for, 2–3, 31, 147

Continental Film Company, 56

Cooper, Jackie, 184

Copyright: violations of, 76

Corbaley, Kate, 93–94, 105, 205n13

Correspondence schools, 2, 4, 5, 13, 19, 20, 68, 70, 80, 81, 86

Corruption, 13, 127, 180

Coughlin, Father, 201n1, 208n15; broadcasting career of, 134; Hays and, 137, 138; and Legion of Decency compared, 135–36

Counts, George, 139

Couvares, Francis, 121, 130, 143; on Catholic Church/Hollywood, 140–41; on clean pictures campaign, 134; on cultural authority, 138; on textual regulation, 116

Cradle Song (movie), 212n22

Craft, 79, 162, 163

Creativity, 79, 82, 85

Criminals, 167, 180, 210n15; crime pictures and, 28; feeble-minded persons and, 10; slums and, 168

Crouch, Matthew, 196, 198
Crowds (Lee), 23, 202n8
Cultural events: masses and, 24–25
Cultural evolution: film appreciation
    movement and, 177
Cultural values, 75, 138
Culture, 3–4, 5, 107; authentic, 108;
    church-sponsored, 206n2; consump-
    tion of, 23; education and, 14, 183;
    entrepreneurial, 193; film, 2, 3, 73, 75,
    105, 195; high, 77, 166; intervention
    of, 183; masculine, 33; mass, 12, 108;
    media, 40; popular, 33, 193, 195;
    religion and, 14; twentieth-century, 41
Cupples and Leon: series fiction and, 35
Cutting script, 102, 205n17
Cvetkovich, Ann: on expression of
    feeling, 67
Czitrom, Daniel: on social environ-
    ment, 212n23

Dale, Edgar, 158, 159, 167, 173, 181; film
    appreciation and, 160, 161; on film-
    making, 175; on hierarchies of
    knowledge, 162–63; NCTE and,
    209n7; Production Code and, 169;
    on social reality, 168
"Debunking" (Motion Picture), 148
DeMille, Cecil B., 73, 88; King of Kings
    and, 122, 202n8; narratives and, 32;
    sex comedies of, 108
Democracy, 19, 153; authority in, 189;
    consumption and, 190; education
    and, 155, 185; film and, 38, 54, 68,
    201n2; future of, 8; industrial, 6–7,
    111; promoting, 79; religion and, 114;
    rhetoric of, 3, 84–85, 97, 201n2
Dennison Manufacturing Co., 166
Department of Child Guidance, 182
Devil Is a Sissy, The (movie), 167
Dewey, John, 162, 187, 212n23; on art,
    183; character education/film
    appreciation and, 155; on culture, 183;
    educational philosophy of, 151, 155,
    156, 157, 160, 183; film/curriculum
    and, 161; group process psychology
    and, 157, 158; on inquiry/social
    action, 158; pedagogy of, 154–59, 186,

189; Progressives and, 139; self-
    realization and, 157; vocational
    education and, 156
Deweyan principles, 150, 165, 175, 186–
    87, 188; application of, 161; Taylorism
    and, 156–57
"Dialectic of Enlightenment, The"
    (Sheridan), 177
Differentiation, 74, 76
Directors: role of, 94, 163
Disney Corporation: protest against,
    198
Distribution, 125, 198; control of, 191;
    Jews and, 129; structure of, 101, 123
Documentaries, homemade, 195
Domesticity, 47, 202n5
"Don'ts and Be Carefuls," 113, 191
Don Ryan's Magazine, 204n2
Doty, Douglas Z., 105
Dougherty, Cardinal, 136
Douglas, Ann, 116, 118, 137
Downes, Bobby, 197
Dramatic Suspense in the Photoplay
    (Clift), 104–5
Drew, Nancy, 35
Dumas, Alexandre, 104, 206n20
Durham, Frank, 37, 203n8
Dyer, Richard: on filmmaking/film
    production, 50
Dynamism, 186–88

"Eastern" writers, 78, 79, 97
East Orange High School: film
    appreciation at, 159
Ecclesiastical Review, 128, 208n19
Economic individualism, 17, 18, 140
Ecumenism, 131, 179
Editorial department, 78
Education: advertising and, 188;
    audience and, 105; citizenship and,
    171; culture and, 14, 183; defects in, 71;
    democracy and, 155, 185; film and, 30,
    113–14, 121, 122, 171; leisure activities
    and, 161; social hygiene and, 171;
    visual, 151. See also Vocational
    education
Educational and Recreational Guides,
    Incorporated, 159

Educational establishment, 190; democracy/authority and, 189; film and, 188–89; moral high ground and, 187; standards and, 188

Educational Pictures, Inc., 206n4

Educational reformers, 14, 28, 31, 152–53, 157; Dewey and, 155; group process psychology and, 133; rhetoric of, 150

Educational Screen, 211n20

Efficiency, 2, 31, 104, 157, 162, 181

Eldridge, Donald: on film appreciation, 164

Elijah, Frances White, 20

Emerson, Alice B. See Fielding, Ruth

Emerson, John: on photoplaywriting, 86

Eminent Authors program, 77, 78, 89, 100, 141

Emotion, 27, 57, 173, 183; appropriate, 64; control of, 68, 176–77; fabricating, 54; laws of, 70; recording, 54; salability of, 67

Emulation, 62, 63

Enfranchisement, 34, 199

Entertainment, 30, 53, 100, 195; family, 28; good/bad, 61; mental age and, 8; middle-class, 28; popular, 25–26, 68, 79; Progressive Era reform of, 47; public and, 192, 199; safe, 2; working-class, 28

Essentials of Photoplay Writing, The, 94

Ewen, Stuart, 19, 21

Exhibition, 28, 125; controlling, 116, 122, 123; Jews and, 129

Exhibitor's Herald, The: Production Code and, 113

Experience, 51, 160, 171

Facey, Paul: on Legion of Decency, 136

Fact: fiction vs., 55–59

Fallen woman cycle, 17

Families: biological/surrogate, 46; capitalism and, 44–45

Famous Players Film Company, 76, 77, 141

Fan magazines, 74, 76, 92, 163

Federal Council of the Churches of Christ in America, 113, 117, 127–28

Fiction, 36–37, 202n3; fact vs., 55–59; personality and, 84; real world of, 69. See also Juvenile series fiction

Fielding, Ruth (character created by Alice B. Emerson), 36, 37, 41, 44, 48, 53, 56, 84, 99; acting by, 49, 51, 57, 58–59, 61, 62, 64–66; business practices of, 62; capital for, 39; complexity of, 20, 65; domesticity and, 47; juvenile series fiction and, 87; longevity of, 41; mobility of, 67; publicity and, 59; screenwriting by, 48, 50, 60, 61, 62; self-commodification and, 59; self-control of, 63–64; series of, 36; work of, 49, 51, 61

Film: accessibility of, 43; controlling, 2, 114, 183; creating, 61; criticism of, 25, 182; defining, 149; historical, 167; independent, 196; industrial, 16, 162; interpretation of, 19; objectionable, 123, 129, 198, 207n13, 207n14; popularity of, 2, 32, 142; power of, 4, 51, 57, 196; psychological workings of, 21–22; quality of, 9–10; religion and, 2, 117, 125, 126, 142, 145, 146, 148, 196

Film appreciation, 3, 28, 33, 57, 149, 150; character education and, 2, 169, 182, 184, 189; curriculum for, 2, 26, 158–60, 165–67, 170, 173, 174, 189; decline of, 193; described, 159–78; education and, 161, 162, 171; film industry and, 162; filmmaking and, 175; film text and, 151; identification and, 166; logical extension of, 174; promoters of, 161; real-life behavior and, 162; religion and, 152; social engineering and, 172–73; social hygiene and, 172–73

Film appreciation movement, 61, 68, 69, 110, 114, 150, 183; autonomy and, 167; censorship and, 163; consumers and, 163; credo of, 165; cultural evolution and, 177; Dewey and, 155; manifesto of, 164; materials of, 165, 167; pedagogy and, 154; standards by, 160; terminology of, 28

Film Booking Office, 103

Film companies, 44; all-woman, 91

Film education movement, 149, 150;
curricula for, 4, 151; standards and,
188

Filmgoers, 14, 75, 78, 119; film industry
and, 17; filmmakers and, 25, 176;
identifying, 105, 118; scenarios and,
139

Filmgoing, 1, 103, 105; democracy and,
68; filmmaking and, 25, 110, 114, 176;
social change and, 152

Film industry, 12, 13, 15, 17, 27, 36–38, 41,
42, 105, 108, 134; accessibility of, 19,
76, 190; character of, 45–46, 193;
control of, 24, 28, 122, 139, 142, 192;
defining, 71, 191; dissatisfaction with,
9, 124; economic structure of, 11–12,
103, 129; flexibility of, 4; good conduct of, 68; information about,
42, 110, 203n12; instruction in, 73;
obstacles by, 190–91; organization
of, 6; Palmer and, 73, 74, 76, 80;
participation in, 109, 190; rhetoric
about, 40, 100; social structures of,
192; technical knowledge of, 75;
training for, 154; understanding, 74;
upgrading by, 77

Filmmaking, 4, 6, 32, 37, 42, 66, 86, 145,
163, 168; aleatory form of, 56; amateur,
174, 193, 195; artistic legitimacy in, 79;
depiction of, 50; difficulties with,
176; extratextual narrative of, 175;
fabricating experience and, 51; fictions
of, 36–37; as industrial production/
art, 76; independent, 195; internal
structure of, 165; involvement in, 37,
55; labor of, 38, 51, 60; ministerial, 3,
117, 127, 195; religion and, 18, 117, 121,
122, 123, 178; series fiction about, 193;
technology of, 66–67; vocabulary of,
82–83

"Film Production in Schools" (Dale),
175

Film Spectator, The, 73

Film, the Democratic Art (Jowett), 201n2

First Amendment, 29, 34, 198

Flaubert, Gustave, 104

Foolish Wives (movie), 107

Ford, Henry, 23, 71

For Husbands Only (movie), 106

Forman, Henry James, 177, 209n5

Formula, 36, 102, 113

Fox, William, 128

Fox Film Corporation, 129, 145

Frankfurt School, 109, 177

Freelance market, 76–77, 86, 87

Freytag, Gustav, 104

Friction, 156, 181

Froebel, Friedrich, 155, 209n3

Frutchey, F. P., 163, 173

Frye, Northrop, of, 67

Fuller, Kathryn, 66; on film industry,
42, 109; on sin and redemption
narratives, 115

"Functions of the Continuity Writer"
(MacPherson), 88

Gallup, George, 11, 147

Gates of Brass (movie), 105

Gender roles, 50, 208n1

General Electric, 154

General Federation of Women's Clubs,
206n6, 207n9

Gentlemen Are Born (movie), 212n22

Ghetto Film School, 193, 194

Golden Bough, The: rights to, 132

Goldwyn, Samuel, 77, 78, 141

Goldwyn Pictures, 89

Gomery, Douglas: on NRA, 120

Good Earth, The (movie), 210n14

Good Will Hunting (movie): discussion
of, 194

Gossman, Lionel: on industrial
capitalism, 15

Gould, Stephen Jay: on test results, 8

Graebner, William, 157, 187

Great Depression, 137, 139, 208n16

Grosset and Dunlap: series fiction
and, 35

Group Discussion Guide, 164, 168,
209n6, 211n18

Group process psychology, 133, 157, 158,
175, 186, 187, 189

Haber, Samuel, 87–88

Hall, Joe: AI and, 194

Halsey, William, 132, 134

Hamburger, Louise: on spiritual freedom, 7

Hammons, Earle W.: Educational Pictures, Inc. and, 206n4

Hanmer, Lee, 14

Hansen, Miriam, 26

Hardy Boys, the, 35

Harmon, William, 125

Harris, Neil, 75, 110

Hawley, Ellis, 120, 153

Hays, Will H., 101, 128, 202n5, 206n5; coalition collapse and, 127; Coughlin and, 137, 138; as Czar of all the Rushes, 124–25; desire of, 102; on production, 148–49

Hays Office, 123, 131, 201n1; Catholics and, 135; censorship and, 207n6; conferences and, 124; democratic opposition and, 124; discrediting, 120–21; film industry and, 120, 126–27; honoraria from, 128; public and, 127, 140; reformers/independent exhibitors and, 125; stars and, 198

Heald, Morrell, 120, 154

Heroes, 37, 38, 39, 54, 68; Algeresque, 47; child, 43; modeling after, 187; as personalities, 27; super, 195

Heroines, 37, 39, 62, 68, 168; active/passive, 61; emotions of, 54; sexual self-commodification by, 108

Higashi, Sumiko: on DeMille, 32

His Forgotten Wife (movie): review of, 99–100, 205n19

Hofstadter, Richard, 24, 133–34

Hollywood: audience and, 79; Catholic Church and, 128, 140–41, 144, 162; cooperating with, 31; education and, 190; openness of, 37; public and, 73; reformers and, 123; religion and, 152, 190

Home Correspondence School, 80, 204n7

Hope, Laura Lee, 36. See Moving Picture Girls

Horace Mann School: film appreciation at, 173, 174

Horkheimer, Max, 40

Houston Press: on Skippy, 184

How to Appreciate Motion Pictures (Dale), 160, 162, 169

How to Judge a Motion Picture (Mullen), 171

How to Write Photoplays (Loos and Emerson), 86

Huckleberry Finn (movie), 212n22

Human behavior, 162; deterministic view of, 151

Huyssen, Andreas: on mass culture/avant-garde, 108

Ibsen, Henrik, 104–5, 206n20

Identification, 24, 37, 166

Immigrants, 8, 24, 115

Ince, Thomas, 73, 205n13

"Increasing General Usefulness of Films" (Motion Picture), 148

Independence, 121–22, 123, 125, 164, 195, 196; rhetoric of, 102

Individual: chance and, 40; film industry and, 3; Palmerplays and, 97–106; reforms/organization and, 153–54

Individuality, 85, 98; assertion of, 110; economic, 152; standardization and, 71, 72

Industrial structures, 46, 120–28, 163

Inness, Sherrie, 42, 53, 62

Insiders, 79, 161

Institute for Propaganda Analysis, 168

Intellectual property, 38, 66

International Federation of Catholic Alumnae, 128, 207n6

International relations: film and, 149

Internet, 192

Intertexts, 32

Jackson, Phyllis, 173

Jacobs, Lea, 2, 149–50, 152, 172; Charters and, 209n5; on film appreciation, 32–33, 166, 182; on plausibility, 167; on Production Code, 29

James, William, 132, 157, 158, 187

Jesus Christ: as business leader, 18

Jews: film industry and, 129–30, 135

Joas, Hans: on Dewey, 183, 212n24; on groups process, 157

Johnson, Sheldon Krag: *Broken Chains* and, 98–99
Journalism: screenwriting and, 90
Jowett, Garth, 124, 201n2, 206n6, 208n16, 209n4
Judgment of the Storm (book by Mason), 103
Judgment of the Storm (movie), 99, 205n17; Manker on, 97; novelization of, 102, 103; praise for, 98
Juvenile series fiction, 2, 4, 7, 13, 16, 19, 22, 26, 33, 35–37, 39, 45–46, 50, 56–57, 68, 72, 76, 84, 87, 100, 114, 146, 150, 152, 176, 190; filmmaking, 193; lessons of, 173; protagonists of, 74; Protestant reformers and, 118–19; publication of, 36; self-commodification and, 5; stars of, 144; women and, 64. *See also Aunt Jane's Nieces Out West;* Fielding, Ruth; Motion Picture Chums; Motion Picture Comrades; Moving Picture Boys; Moving Picture Girls

Katterjohn, Monte, 80
Keene, John H.: on "Secrets of Success," 178
Kendall, Miss, 91
Keyser, Barbara, 143, 144
Keyser, Les, 143, 144
Kimball, Winifred, 92; approval for, 98–99; photoplay and, 89–90, 91
King of Kings, The (movie), 117, 202n8; Christ representation in, 142; profits for, 122, 127
Klein, Marcus: on Alger, 39
Knights of Columbus: protest by, 198
Knowledge: extratextual, 192; hierarchies of, 162–63; technical, 75
Koszarski, Richard, 78
Krows, Arthur Edwin, 9, 80
Ku Klux Klan, 134
Kulturkampf, 138
Kummer, Frederick Arnold, 86, 94

Lane, Tamar, 7–8, 15, 100, 101
Lasky, Jesse, 77
Leach, Eugene: on conception of crowds, 23

Learning: arousal and, 183, 184
Lears, T. J. Jackson, 84, 158, 201n3
Lee, Gerald Stanley, 22, 202n8
Legion of Decency, 2, 24, 137, 205n16, 207n13; campaign by, 128–29, 134, 135, 140, 141, 201n1; Catholics and, 130–31; censorship and, 140; corporate menace and, 146; and Coughlin compared, 135–36; film industry and, 138–39; pledge of, 117; populist revolt and, 134, 139; public and, 140, 146; rating system and, 131; structure of, 136
Leisure, 20, 22, 161; right use of, 28, 187
LeSourd, Howard M., 179, 181, 182; on compelling instructors, 184; dynamism and, 186, 187; *Secrets of Success* and, 187; on standardization, 185
"Let's Talk It Over," 182
Letter-writing campaigns, 139
Levine, Lawrence, 24, 25
Lewin, William, 21, 210nn10, 16, 211n18; on consumers, 166; on education/democracy, 185; Educational and Recreational Guides and, 159; film appreciation and, 159, 161, 162, 165, 171; film industry and, 209–10n9; NCTE and, 160; on photoplays, 169; on social reality, 168
Lewis, Sinclair, 70
Life of Moses, The (movie), 117
Lifestyles: commodification of, 18–19
Lippmann, Walter, 12, 13, 137
Literary Digest, 108
Literature, 16, 105; movies and, 9; photoplays and, 171; Progressive Era, 85; women and, 116
"Little Men behind the Big Screen, The," 13
Little Screen Players of Boston, 211n21
Little Stories of Success (Palmer), 87, 106
Locations: finding, 58
Loos, Anita: on photoplaywriting, 86
Lord, Daniel, 117, 132, 144, 208n18; cultural mores and, 131; Production Code and, 113, 143
Lord, Martha: on *Palmer Photoplay Encyclopedia,* 104

Lowenthal, Leo, 49, 65
Lower West Side Motion Picture
    Council, 181
Lucky Dog (movie), 212n22
Lynd, Helen, 48, 115
Lynd, Robert, 48, 115

Macmillan: film appreciation and, 160
MacPherson, Jeanie, 88
Magoun, F. A., 184; on learning/arousal,
    183; on "Secrets of Success," 182–83
Maid of Salem (movie), 210n14
Maltby, Richard, 108, 122, 202n9, 209n6;
    on Catholic Church/Hollywood,
    140–41; on Hays, 101–2, 120–21; on
    Production Code, 29, 133; on
    religion/filmmaking, 121; on rural
    Protestants, 115; on self-fetishization,
    17; on sin and redemption narratives,
    115; on study guides, 165
Management, 37; children in, 47; laws
    of, 88
Manhattan Motion Picture Institute, 80
Manipulation, 37, 65, 67
Manker, Roy, 205n15; concerns of, 107–
    8; on *Judgment of the Storm*, 97; on
    Palmer/women, 106; personality and,
    83–84; Ryan and, 91; on screenwriting,
    90; on woman's picture, 106–7
Man Nobody Knows, The (Barton), 18,
    202n8
Manufacturing, 28, 45, 48, 74, 100
Manuscripts: critiquing, 80; mass-
    produced, 94, 96
Marketplace, 11, 51, 62, 116, 118
Market research, 66, 93, 158
Marston, William, 146, 147
Marx, Samuel: on Corbaley, 205n13
Mason, Roy, 103
Mass: self and, 22–25
Materialism, 62, 134
May, Lary: on stars, 64–65
Mayer, Louis B., 141, 205n13
Mayne, Judith, 23
McGoldrick, Mrs. Thomas A., 128
McPherson, Aimee Semple: film
    company of, 117
Media, 12, 34, 40, 179, 185

Megiddo (movie), 196
Melodrama, 44, 72, 83, 98, 99, 100, 173,
    176, 177
Mental age, 8, 9, 10
Mental punch: physical punch and, 99
Men with Wings (movie), 164
Mercy Streets (movie), 197
Methodists: filmmaking and, 118
Metro-Goldwyn-Mayer (MGM), 93,
    205n13
Michaels, Walter Benn: on Progressive
    Era literature, 85
Middleton, Ethel Styles, 98, 103
Middletown in Transition (Lynd and
    Lynd), 48
"Minister as a Movie Maker, A"
    (Coffin), 119
Miracle Man, The (movie), 105
Modernism, 132, 206n20
Moley, Raymond, 123, 131, 211n17
Mones, Leon: whole impression school
    and, 171
Montclair High School: filmmaking
    at, 175
Moon, Michael, 39; on Alger, 40–41,
    42, 63
Moral issues, 11, 22, 28, 48, 149, 182, 183–
    84, 191; addressing, 170, 186; censors
    and, 8; citizenship and, 19; cultural,
    131; erosion of, 16–17; film and, 64,
    121, 122, 130; filmmaking and, 43;
    heroes for, 187
Motion Picture, The, 146, 148
Motion Picture and the Family, The, 167,
    178, 179, 211n22; production of, 159
Motion Picture Chums (series by Elmer
    Tracey Barnes), 36, 37, 38, 203n8;
    family of, 45, 46, 47; housekeeping
    and, 203n6; National Theater and, 53;
    success of, 40, 41; Wonderland and,
    52–53
Motion Picture Classic, 206n21
Motion Picture Comrades (series by
    Elmer Tracey Barnes), 36
Motion Picture Council, 181
Motion Picture Magazine, 101, 206n21
Motion Picture Monthly, The, 149, 154;
    on *Skippy*, 184

Motion Picture News, MPPDA and, 207n8

Motion Picture Patents Company (MPPC), 29

Motion Picture Producers and Distributors of America (MPPDA), 14, 92, 110, 113, 120, 121, 127, 130, 134, 168; *Motion Picture News* and, 207n8; organ of, 146, 148; reformist opposition and, 124; screenwriting and, 77; sponsorship by, 111; struggle with, 116; voluntary organizations and, 153–54; Wallace and, 125–26

Motion Picture Research Council, 209n5

Motion Pictures and Standards of Morality (Peters), 19

Motion Pictures in Education: Dale and, 211n20

Motion-Picture Study Groups (Pollard), 210n11

Motion Picture Theatre Owners of America, 125

Movie Makers, 211n20

"Movies: A Colossus That Totters, The" *(Bookman),* 78

Movies, Delinquency, and Crime (Blumer and Hauser), 208n19

Moving Picture Boys (characters created by Victor Appleton), 36, 51–53, 58–59; advertising of, 66; autonomy of, 43; nonfiction events and, 43, 50, 51; shooting by, 203n10

Moving Picture Girls (characters created by Laura Lee Hope), 36, 37, 45, 47–49, 51–54, 57, 61

Moving Picture World, 96, 107

MPPC. *See* Motion Picture Patents Company

Mullen, Sarah McLean, 17, 209n7, 210n12

Mundelein, Cardinal, 136, 208n17

Music for Madame (movie), 171

Mutiny on the Bounty (movie), 166

Narratives, 15, 16, 23, 31, 44, 46, 64, 68, 79, 100, 104, 107, 109; Algeresque, 41; development of, 26, 29, 175;

embedded, 66; fostering, 83–84, 176; girls and, 48; matter/modes of, 32; nonfiction, 43; protagonists of, 37. *See also* Fiction; Juvenile series fiction

National Board of Censorship, 29

National Board of Review, 29, 30–31, 129–30, 145, 181; artistic experimentation and, 132; censorship and, 164; conferences and, 124; formation of, 153; permissiveness of, 115

National Catholic Welfare Congress, 207n6

National Congress of Parents and Teachers, 207nn6, 9

National Council of Teachers of English (NCTE), 159, 160, 192, 209nn6, 7, 211n18

National Education Association, 207n6

National Recovery Administration (NRA), 120

National Theater: Motion Picture Chums and, 53

National Union for Social Justice, 136

Natural law, 133, 141, 143, 144, 189; defying, 169

NCTE. *See* National Council of Teachers of English

Nelson, Richard Alan, 117

Neo-Thomism, 133

New Era, 140

Newsweek: on ministerial filmmaking, 195

New York City Public Schools, 182

New York Times, 210n12

Nicolosi, Barbara, 197

Nineteenth Amendment, 24

No Greater Glory (movie), 212n22

"Noted Teacher" (Magoun), 182–83

Novelization, 102, 103

NRA. *See* National Recovery Administration

O'Brien, Pat, 143

O'Higgins, Harvey, 88

Ohmer, Susan, 11

"Oliver Optic" tales, 39

Omega Code, The (movie), 196, 197

Originality, 79
Our Movie-Made Children (Forman),
    177, 209n5
Outsiders, 1, 2, 3, 79, 166

Pacifism: film used to inculcate, 149
Palmer, Frederick, 71, 95, 197, 203n1,
    204n2; freelance market and, 87;
    Hays Office and, 73; as mentor/muse,
    80–81; Palmerplays and, 92; on
    physical/mental punch, 99; produc-
    tion and, 97; scenarios and, 81, 97,
    205n11; successes for, 96; women's
    scripts and, 107; on writing/selling
    stories, 105
Palmer Clubs, 95
Palmer Institute of Authorship, 205n13
Palmer Photoplay Corporation, 3, 4, 7,
    16, 20, 22, 24, 26, 33, 37–38, 41, 42, 48,
    61, 73–74, 84, 94, 95, 111, 114; adver-
    tising by, 71, 72, 82, 85, 86, 88, 125,
    144, 194–95, 205n10; applying to, 97;
    audience and, 118–19; business of, 71;
    curriculum of, 81–83, 98; empire of,
    79–81; founding of, 70; individuality
    and, 85; literature of, 14, 75–76,
    152, 172, 194, 204n5; rhetoric of, 77,
    89, 150, 151; rivals for, 80; self-
    commodification and, 5, 68; vision
    of, 38
Palmer Photoplay Encyclopedia, 104
Palmer Photoplay School: His Forgotten
    Wife and, 205n19
Palmer Plan, 86, 87, 89, 95, 110–11, 197;
    appeal of, 83–84; filmgoing/filmgoers
    and, 105; products of, 93, 106;
    screenwriting and, 92–93
Palmerplays, 80; first, 92, 97, 98;
    individual and, 97–106; reviews of,
    99–100
Paramount, 129
Participation, 7, 24, 72, 152, 164, 190, 191;
    amateur, 109, 174; illusion of, 34;
    Palmer and, 139; public, 89, 109, 112,
    140, 193; rank-and-file, 180; rhetoric
    of, 193
Patterson, Frances Taylor, 12, 80,
    201–2n4

Payne Fund studies, 158, 167, 177,
    209nn5, 7, 11; Dewey and, 188; film
    appreciation and, 160
PCA. See Production Code
    Administration
Pearson, Roberta: on Vitagraph, 117
Pedagogy, 154–59, 186, 189; principles
    of, 182
Pells, Richard: on Dewey, 139
People's Institute, 29
Performance, 44, 66; control over, 25,
    65; involuntary, 55; stealing, 55
Personality, 20, 83–84, 86–87; commod-
    ification of, 18–19; culture of, 85;
    defects in, 71; emphasis on, 84; fiction
    writing and, 84; hero as, 27; produc-
    tion of, 100; transformation of, 195
Pestalozzi, Johann, 155
Peters, Charles, 19
Photo-Drama of Creation, 117
Photodramatist (house organ), 81, 89, 95,
    105, 204n5; on applicants, 96; reading
    of, 101; on scenario writing, 80
Photodramatist, The (house organ),
    204n5, 205n15
Photography, 52, 94, 170
Photoplay, 81; advertising in, 76, 86;
    articles in, 92; Loos/Emerson in, 86
Photoplay Appreciation (Lewin), 211n18
Photoplay Club, 165, 166, 173, 210n12
Photoplay Enterprise Association, 80
Photoplay Plot Encyclopedia, The, 79,
    104, 194, 204n7
Photoplays, 94, 96, 104; appraising, 160;
    collaboration in, 95; literature and,
    171; selecting/judging, 169; story and,
    102–3; success in, 88; understanding,
    82; writing, 72, 97, 197
Photoplay Studies, 209n6, 210n13
Photoplaywright's Handy Text-Book,
    The (Radinoff), 80
Photoplaywright, The (house organ),
    20, 96, 204n5
Physical punch: mental punch and, 99
Pickford, Mary, 65
Pinkham, Lydia, 7
Piracy: problems with, 38, 57–61, 78–79.
    See also Plagiarism

Placement department, 80
Plagiarism, 77, 93, 94. *See also* Piracy
Plainsman, The (movie), 210n14
Plausibility, 83, 167, 168
Plots, 94, 104; mystery, 54, 57
Point of Attack, or How to Start the Photoplay, The (Badger), 104
Pollard, Elizabeth Watson, 172, 210n11
Pragmatism, 133, 157
Pre-sold properties, 73, 103
Presbyterian, The: on *Skippy*, 184
Pressure groups, 124, 191, 198
Priest (movie): boycott of, 198
Priests: fighting, 143; homosexual, 198
Prince and the Pauper, The (movie), 210n14
Producers:, audience and, 71, 114
Production, 16, 100, 101, 162; amateur, 110; capital behind, 13–14; consumption and, 172; control over, 5, 191; costs of, 122; efficiency in, 31; factory-inspired methods of, 35; filmmaking and, 50; idols of, 49; industrial, 97; information on, 165; Jews and, 129; layers of, 27; mass, 18, 151; reformism and, 27–31; resurgent, 198; social studies and, 174; technology of, 110; urban/immigrant culture of, 115; values, 197; video, 195–96
Production Code, 2, 3, 10, 12, 13, 16, 24, 27, 31, 110, 111; amateur, 174; Catholic Church and, 11, 169, 189; corporate menace and, 146; enforcement of, 112, 113, 120, 121, 129, 189; establishment of, 11, 135, 191; on film, 32, 176; installation of, 192; justification for, 131, 132; philosophical presentation of, 143; public and, 146; schemes, 92, 103; struggles over, 113, 114, 141
Production Code Administration (PCA), 30, 121; Catholic Church and, 28, 113; enforcement by, 112–13, 135; success of, 29
Production companies, 41, 49
Progressive Era, 23, 30, 46, 140, 163; Dewey and, 139; individualism and, 153; reform during, 47; Taylorism and, 88

"Proposed Community Motion Picture Program, A," 181
Protagonists, 37, 38, 52, 53, 119; active/passive, 61; feelings of, 68; series fiction, 74
Protestant Church, 17, 115; opposition from, 29
Protestant reformers, 134; artistic experimentation and, 132; authorship and, 116–20; Catholics and, 24; film industry and, 114, 115, 119; filmmaking and, 116, 117–18; juvenile series fiction and, 118–19
Public, 23, 120; film and, 31–34; film industry and, 1, 27–28, 28, 76, 124; scriptwriting and, 146
Public opinion, 7–14, 118
Public relations, 124, 127, 138
Public Relations Conference of the Motion Picture Industry, 179
Public Relations of the Motion Picture Industry, The, 113
Public service, 124, 128, 146, 207n9
Publishers Weekly, 35
Publishing: Taylorism and, 35; and film industry compared, 101
Pumphrey, Martin: on women's magazines, 106

Queen's Work, The (movie), 131, 208n18
Quigley, Martin, 113, 136, 143, 189, 208n18

Radinoff, Florence, 80
Radio Boys (series by Allen Chapman), 36
Rand, Helen Miller, 167, 168
Rating scales, 131, 169–70, 171
Ray, Charles, 86
Reality, 99, 186–87; detachment from, 13
Reception, 26, 28, 191
Reform, 153–54; attempts at, 6; Christian, 4; discourses of, 22; forces of, 29; ideal approach for, 14; production and, 27–31; Progressive Era, 47
Reformers, 122, 153; corruption among, 127; film industry and, 18, 123, 126–27, 191; masses and, 24; religious, 28;

Reformers *(continued)*, textual system and, 27; total picture school of, 32. *See also* Catholic reformers; Educational reformers; Protestant reformers

"Reformers and Spectators: The Film Education Movement in the Thirties" (Jacobs), 2, 149–50, 172

Reilly and Britton, 36

"Rejected One, The" (Kummer), 86

Religion: box office and, 135; culture and, 14; democracy and, 114; film and, 2, 145, 146, 148; film industry and, 113–15, 119; filmmaking and, 18, 117, 121, 122, 123, 178; Hollywood and, 152, 190; social adjustment and, 146; social life and, 145

Religious Motion Picture Foundation, 125

Repression: rise and, 63–67

Rhetoric, 77, 79, 86, 89, 95, 100, 101, 103, 113, 114; anti-Semitic, 129–30; box office, 21; contradictions in, 72; corporate service, 154; democracy, 3, 84–85, 97, 201n2; electoral, 127; film, 21, 40; independence, 102; participation, 193; self-commodification, 187; service, 154

Richardson, Samuel, 23

Rinehart, Mary Roberts, 78

Rise: repression and, 63–67

Roosevelt, Franklin D., 134

Rosenzweig, Roy, 31

Rover Boys (series by Arthur M. Winfield), 35

Russell, Charles Taze, 117

Ryan, Don, 90, 91, 92, 204n2

Sales department, 94; plagiarism and, 93

Sargent, Epes Winthrop, 92, 103

Scenario department: piracy/bias and, 78

Scenario writing, 80, 94, 97, 195; filmgoers and, 139; yin/yang of, 85

Scholastic Magazine, 211n18

School and Society, The (Dewey), 209n3

Schultheiss, John: on entrenched bureaucracy, 78

Schwarzenegger, Arnold, 196

Science: film and, 148

Screen, The: Lane and, 101

Screenwriting, 48, 71, 72, 82, 88, 89, 91, 111, 195; acceptable, 1–2; advice on, 92–93; amateur, 79; craft of, 110; industrial context of, 73–79; interest in, 81; journalism and, 90; non-academic instruction in, 194; self-commodification and, 87; teaching, 80, 81; women and, 50, 106–7

Scripts, 102; standardization of, 74; submitting, 93; vetting, 112–13

Scriptwriting, 5, 146, 204n6; amateur 109

"Secrets of Success" (film series), 178–79, 181, 184, 186, 187; praise for, 180, 182–83

Selection programs, 122, 163–64

Self: alienation of, 87; film and, 25–27; intellectual property and, 66; mass and, 22–25; producing/controlling/ selling, 17–18

Self-censorship, 29, 164, 168

Self-command, 33, 65

Self-commodification, 19, 27, 32, 36, 39, 62, 152; advertising and, 83–92; extreme, 59; individual, 114; institutional, 114; Palmer and, 5, 68, 83; price of, 68; process of, 74; project of, 84; rhetoric of, 87, 187; screenwriting and, 87; sexual, 17, 108

Self-control, 63–64, 67

Self-cultivation, 21, 49, 50–54, 63

Self-discipline, 44–45, 63, 68, 69, 175

Self-expression, 5, 51, 52, 71, 98; individual/group, 193; Taylorization of, 72, 109; tools for, 83

Self-fashioning, 34, 146

Self-mastery, 57, 63

Self-regulation, 9, 17, 120, 154; industry, 114; strictures of, 29–30

Self-revelation, 50–54

Sensation seekers, 172

Sentimentalism, 8, 168

Sex stimulation, 8, 107–8, 115, 144, 208n19

Shakespeare, William, 24, 25

Sheridan, Marion: on rescuing civilization, 177

Short, William H., 188, 209n5

Shute, Mrs., 91
Sign, 208n19
Sign of the Cross, The (movie), 212n22
Sin and redemption: rhetoric of, 115
Skill: filmic, 79; performance and, 66; problem-solving, 156, 157; professional, 5
Skinner, R. Dana, 142
Skippy (movie), 184, 186, 212n22
Smith, Al, 133
Smith, Courtland, 125–26
Smoodin, Eric, 151, 208n1
Social activism, 158; Catholic Church and, 133, 134, 145
Social agencies, 144, 179
Social change, 17; film and, 149; movie-going and, 152; religion and, 146
Social engineering, 151, 157; film and, 148, 149, 172–73
Social hygiene, 171, 172–73
Socializing goals, 184
Social justice, 134
Social life, 145, 162
Social order, 155–56, 157
Social power, 33, 110, 156
Social problems, 30, 67, 157, 174; discussion of, 115; filmic discussion of, 113; resolution of, 175
Social reality, 152, 168–69
Social rise, 7, 48, 67, 140; affect and, 42–44; barriers against, 40; individual, 68; rhetoric of, 174
"Social Values in Motion Pictures," 179
Social workers, 154, 181, 182, 194
Social Work Year Book 1929, 14
Society: film and, 32, 110; mass, 22, 23
Sooky (music), 186
Special interests, 128, 134, 201n1
Spectacle, 75, 167
Springer, John Parris, 39, 202n1
Staiger, Janet, 74, 211n19; on culture/filmmaking, 75; on freelance market, 76; on woman's picture, 107
Standardization, 16, 18, 83, 85, 88, 136, 152, 169, 186; differentiation and, 76; film industry and, 153; goals of, 74; individuality and, 71, 72; overreliance on, 185; soulless, 123–24

Standards, 172; commodification of, 18–19; developing, 169, 173, 188; film, 14, 79, 153, 188, 190; incorporation of, 69; industrial/professional, 190; internationalization of, 5; uniformity in, 170
Stars, 33, 68, 163; construction of, 176; film industry and, 64–65; juvenile, 144; literature about, 76; manipulation of, 175; pressure on, 139; publicity and, 138; scandals involving, 198
"Stenographic Report of Character Education Lesson," 186
Stereotypes, 7, 12, 143
Stockton, W. H.: filmmaking by, 119
Stokes, Cecil A.: on experimentation, 118
Stories: commissioning of, 76; constructing, 75; in-house, 77; original, 72, 73; photoplay and, 102–3; reworking, 75
StoryCraft, 194
Story of Film, The (Hammons), 206n4
Storytelling, 58, 75, 167
Story World and Photodramatist, The (house organ), 91, 103, 204n5
Stratemeyer, Edward, 39, 40, 202n4, 203n8; children of, 43; series fiction and, 35; technology and, 41, 42
Street and Smith, 39
Studio Relations Committee, 121
Studlar, Gaylyn, 33, 107
"Study Your Audience" (O'Higgins), 88
Submissions, 93–94, 112–13
Sum-of-the-individual-parts school, 171, 211n19
Susman, Warren, 19–20, 84
Swift, Tom (series by Victor Appleton), 36

Taste, 133; children's, 173; development of, 160, 169, 192; educated, 172; film industry and, 9; gauging, 146, 168; reliability of, 172
Taurog, Norma, 184
Taylor, Frederick Winslow, 15, 18, 23, 108, 111, 181; bonus system of, 72; efficiency and, 157; friction and, 156; management and, 88; production and, 16, 202n7

Taylorism, 14–17, 72, 83, 86, 95, 111, 150, 155, 158; Deweyism and, 156–57; film and, 23–24; Progressives and, 88; publishing and, 35; self-expression and, 109

Teachers: authority for, 188; extraordinary, 185; poorly prepared, 184, 185; standardized roles for, 186

Teacher's Key to Accompany the Student's Guide to Critical Appreciation of the Photoplay Version of Stevenson's "Treasure Island," 164

Technique of the Photoplay (Sargent), 92

Technocracy, 13, 89

Technology, 14, 41, 82–83, 175, 193; film, 42, 44; problematic, 44; production, 110; screenwriting, 205n17; social dislocation and, 181; using, 75

Television, 192, 196

"Ten-Point Film Program," 164

Terman, Louis: on feeble-minded persons/criminals, 10

Texts: audience and, 27; control over, 26, 31; filmic, 15, 31, 32, 57, 141, 151, 167, 173, 186–87; mastery of, 32; plausible/truthful, 167; reality and, 186–87; shaping, 141, 150; systems, 26–27

Thank U (music): making of, 145

Theaters: independent, 122, 123, 125, 164; local, 164–65; quasi-independent, 121–22; Sabbath operation of, 117; studio-owned, 121

Their Adopted Country (movie), 174, 175, 176

There's Always Tomorrow (movie), 212n22

"Thirteen Points," 113

Thompson, J. Walter, 187

Thompson, Kristin: on classical cinema, 26

Thought, 208n19

Thrasher, Frederic, 180–81

Tichi, Cecelia, 156–57, 204n10

Toast of New York, The (movie), 210n14

Tom Brown of Culver (movie), 212n22

Tom Sawyer (movie), 212n22

Tracy, Spencer, 143

Trade press, 74, 77, 207n11

Triangle Shirtwaist Company, 45, 47

Universal Pictures, 146

Universal Plot Catalogue, The (Philips), 204n7

Uricchio, William: on Vitagraph, 117

Valentino, Rudolph, 107, 206n21

Van Dyne, Edith. *See Aunt Jane's Nieces Out West*

Variety, 11, 77, 103; on Methodists/filmmaking, 118; review in, 98, 99–100, 205n19

Vasey, Ruth, 127, 207n6

Vaughn, Stephen, 135

Victoria the Great (movie), 166

Video cameras, 192

Visualization, 81–82

Vocational education, 156, 174, 175, 210n9; film appreciation and, 161, 162

Von Stroheim, Erich, 107

Vulgarity, 8, 9, 17, 173, 191

Wagner, Rob: on adaptations, 101

Wallace, Lew, Jr., 207n8; Hays Office and, 126; MPPDA and, 125–26

Wal-Mart: distribution by, 198

Walsh, Francis Augustine, 133, 143, 208n18; on Legion of Decency, 136

Wanamaker's Department Store, 188

Ward, John William: on individualism, 85

Way Down East (movie), 99

WCTU. *See* Women's Christian Temperance Union

"We Are All Censors" (Boyle), 140

Weber, Lois, 73, 106

Wednesday's Child (movie), 212n22

Weequahic High School, 210n12

Westbrook, Robert: on study of education, 155

"What Is Appreciative Inquiry?" (Hall), 194

What's Wrong with the Movies? (Lane), 7, 15

White Sin, The (movie): review of, 99
Whole impression school, 171, 211n19
"Why I am Ashamed of the Movies"
  (Collier's), 9
Wiebe, Robert: on national progres-
  sivism, 153
Will to Believe, The (James): rights to, 132
Winter, Alice Ames: on Palmer Plan,
  110–11
Woman's picture, 106–7
Women: film and, 50, 51, 70, 91; literary
  marketplace and, 116; Palmer and,
  106–9; screenwriting and, 50, 106–7;
  series fiction and, 64

Women's Christian Temperance Union
  (WCTU), 30, 207n6
Women's Municipal League, 29
World Court, 134
Writer's Monthly, 80, 86

Yost, Edna, 35
"You'll Get What You Ask For"
  (Collier's), 10–11
Young, Donald, 18, 27
Young America (movie), 211n22

Zimmermann, Patricia, 174, 206n3
Zook, George F., 180
Zukor, Adolph, 9, 128, 141

Anne Morey is assistant professor of English and performance studies at Texas A&M University.